# LIFELINE EARTHQUAKE ENGINEERING IN THE CENTRAL AND EASTERN U.S.

Proceedings of three sessions sponsored by the Technical Council on Lifeline Earthquake Engineering in conjunction with the ASCE National Convention in New York, New York, September 1992

**Technical Council on Lifeline Earthquake Engineering**
**Monograph No. 5**
**September, 1992**

**Edited by Donald B. Ballantyne**

Published by the
American Society of Civil Engineers
345 East 47th Street
New York, New York 10017-2398

**ABSTRACT**

This proceedings, *Lifeline Earthquake Engineering in the Central and Eastern U.S.*, is a collection of papers presented at the 1992 American Society of Civil Engineers International Convention & Exposition in New York, New York, September 14-17, 1992. The papers are grouped into three categories: water supply, transportation systems, and energy distribution. Earthquake hazards and potential losses are discussed on both regional and local levels. Many of the studies focus on the New Madrid seismic zone. Mitigation of damage from major earthquakes with long recurrence intervals pose different types of problems than those encountered in the west. Environmental impact and associated costs are discussed for seismically damaged oil transmission lines. Mitigation approaches are presented for water systems and oil pipelines.

**Library of Congress Cataloging-in-Publication Data**

ASCE National Convention (1992: New York, N.Y.)
Lifeline earthquake engineering in the central and eastern U.S.: proceedings of three sessions sponsored by the Technical Council on Lifeline Earthquake Engineering Research at the ASCE National Convention in New York, New York, September 1992/edited by Donald B. Ballantyne.
p. cm. — (Monograph/Technical Council on Lifeline Earthquake Engineering: no. 5)
Includes index.
ISBN 0-87262-902-3
1. Lifeline earthquake engineering—United States—Congresses. 2. Public utilities—United States—Earthquake effects—Congresses. 3. Transportation — United States — Earthquake effects—Congresses. 4. Pipelines—United States—Earthquake effects — Congresses. I. Ballantyne, Donald B. II. American Society of Civil Engineers. Technical Council on Lifeline Earthquake Engineering. III. Title. IV. Series: Monograph (American Society of Civil Engineers. Technical Council on Lifeline Earthquake Engineering): no. 5.
TA654.6.A85 1992
624.1'762—dc20 92-26364
CIP




Library of Congress Catalog Card No: 92-26364
ISBN 0-87262-902-3
Manufactured in the United States of America.

# PREFACE

The Central United States New Madrid source zone produced one of largest earthquakes known to man. Few people saw its effects. Today, the same area is heavily developed; the earthquake impact would be devastating. Other earthquakes occurred prior to the turn of the century in areas such as Charleston, South Carolina and Boston. Recurrence today would also be catastrophic. Earthquakes with long recurrence periods such as in the Central and Eastern U.S. have not driven designers to incorporate earthquake provisions into design. Therefore, while the probability of having an event is small, the potential risk to lifeline facilities is very large.

One of the most serious potential consequences of a Central U.S. earthquake is the impact on energy distribution to areas far removed from the epicenter. The Mississippi Valley corridor carries much of the oil and natural gas from the Gulf of Mexico to the Northeast. Damage to oil transmission lines along that corridor could, for example, have a long term impact on home heating and industrial activity in Buffalo. Water supply and transportation systems, in addition to energy distribution, are fundamental elements of our nation's infrastructure supporting society. Earthquake damage could be extremely disruptive. This Monograph focuses on those three lifelines.

In 1991, the Technical Council on Lifeline Earthquake Engineering, TCLEE, sponsored the Third U.S. Conference on Lifeline Earthquake Engineering in Los Angeles, 20 years after the San Fernando earthquake. The spotlight was on the west coast.

The objective of this Monograph is to allow readers to focus on earthquake risk in the Central and Eastern United States. There is an increasing level of interest and technical activity in lifeline earthquake engineering east of the Rockies. The National Center for Earthquake Engineering Research, NCEER, is the basis for much of that effort. NCEER was the joint sponsor of the three conference sessions with TCLEE. The conference proceedings resulted in this document. The three conference session coordinators who were instrumental in pulling together the papers for their particular sessions each have major involvement in NCEER as follows:

- Energy Distribution Session—Professor Masanobu Shinozuka, Director, NCEER
- Transportation Systems Session—Dr. Ian Buckle, Deputy Director, NCEER
- Water Supply Session—Professor Michael O'Rourke, Rensselaer Polytechnic Institute, Principal Investigator for many NCEER projects

Their assistance in that effort is greatly appreciated.

Each of the papers included in this Monograph has been accepted for publication by the Monograph Editor. TCLEE Monographs are not published regularly, and there is no provision for discussion. All papers are eligible for ASCE awards.

Donald B. Ballantyne
Editor

# CONTENTS

## Water Supply



## Transportation Systems



## Energy Distribution

# EARTHQUAKE HAZARD INVESTIGATIVE PROCEDURES FOR CENTRAL UNITED STATES WATERWORKS

James R. Blacklock[1]

## Abstract

The vulnerability assessment of Central United States waterworks for earthquake hazards is a developing engineering practice for civil engineers in Arkansas, Missouri and states close to the New Madrid Fault. Because of a lack of seismic experience and an approved regional vulnerability assessment methodology, many recent preliminary engineering inspections and evaluations of lifeline facilities have been performed by out-of-state earthquake engineering consultants. Because of heightened seismic awareness and new seismic building code laws in several states in the New Madrid Fault region, the decade of the 90s will be important for regional Civil Engineers to undertake the tasks of performing seismic vulnerability assessments for utilities and lifelines. This paper addresses some of the issues of civil engineering practice for this important work of assessing risks for lifeline systems in the New Madrid Earthquake states.

## Introduction

The low frequency of major earthquakes in the Central United States has allowed the current generation of practicing waterworks civil engineers in the region to avoid direct involvement with a damaging earthquake event. As a result, most waterworks facilities in Arkansas, Missouri and the other New Madrid states were not constructed seismically-resistant. The most active portions of the New Madrid fault are in Arkansas and Missouri. The southwest portion of the fault is located in northeast Arkansas and the northwest portion is located in southeast Missouri. Although the four great New Madrid Earthquakes that occurred in Arkansas and Missouri in 1811 and 1812 constituted the largest single series of earthquakes ever recorded, they occurred so many years ago, when native American Indians

----------------

[1] Professor of Construction Engineering Technology, University of Arkansas at Little Rock, 2801 South University Avenue, Little Rock, Arkansas, 72204.

outnumbered settlers, that they almost had been forgotten until recent time. Now, largely because of the early work of the late Otto Nuttli at St. Louis University (Nuttli, 1990), and the recent federally funded preparedness programs, there now exists a sizeable awareness that the earthquake risk in the New Madrid Seismic Zone (NMSZ) ranks it as one of the major risk areas in the United States.

As a result of this recent awareness and exposure to public television coverage of major earthquakes of the 80s, many of the major small city utility companies in these states have recently begun serious earthquake hazard reduction mitigation programs. This paper discusses earthquake hazard investigative procedures for waterworks facilities in areas of large to moderate infrequent earthquake events such as Arkansas, Missouri and the other central states.

In the past few years, several Arkansas and Missouri cities, with urging from the state offices of emergency services, have begun conducting seismic hazard investigations and implementing hazard mitigation programs. One important element of these programs has been the preliminary seismic hazard investigation. This paper discusses the steps involved in a waterworks facility preliminary seismic hazards investigation. Standard seismic hazard investigation procedures developed on the west coast have been extended to the Central United States to address the construction types common to the New Madrid Seismic Zone. Very few waterworks buildings in the Central United States have been constructed according to the provisions of any standard seismic building code. Most buildings, even modern buildings, have been constructed with unreinforced masonry bearing walls with poorly attached weak roof and ceiling diaphragms. The foundation soils in the Mississippi River valley are mostly unconsolidated alluvium situated in deposits that are often unstable or easily liquefiable (Fuller, 1912).

The shortage of seismic instrumentation data and the technology necessary to calculate seismic attenuation rates and ground motions in this region has assumed a new level of importance to local civil engineers involved in seismic studies of lifeline facilities.

## Developing Technology

The techniques for inspecting waterworks buildings and structures in Arkansas and Missouri must by necessity be part of a developing technology. As new information concerning predicted earthquake attenuation, recurrence and magnitude levels is presented for the New Madrid fault, earthquake engineering investigators should work to upgrade their technical methods and procedures. With the new seismic attenuation, recurrence, and ground motion information that is promised for the future, it should be possible to calculate the much-needed site-specific, event-specific response spectra for most major small city population centers near the fault. Without better information, lifeline risk evaluations for specific sites must be based on the assessment of existing technical data concerning seismic hazards that provide scant particulars.

In general, Arkansas and Missouri buildings are aseismically designed and constructed, making it easy and tempting for seismic evaluators to make negative generalized statements concerning the perceived seismic risk. Without engineering calculations, these assessments are preliminary at best, lacking the necessary rigor needed to support large expensive retrofit and mitigation programs.

Recent seismic waterworks vulnerability assessments in Arkansas, Missouri and the other central states concerned with future seismic events on the New Madrid fault, have been mostly locally funded and have resulted in limited documentation. Very little published information is available concerning the procedures that have been utilized. The exceptions are several nationally funded waterworks studies concerning Memphis, Tennessee, which have resulted in the following valuable publications (Hwang,1991; Hwang and Lee, 1990; and O'Rourke and Russell, 1991), which utilize seismic data for a probable event in the southern extreme of the NMSZ located in northeastern Arkansas. Also, a recent FEMA study for St. Louis, Missouri, "Estimated Future Earthquake Losses For St. Louis City and County, Missouri," June, 1990, offers valuable predictions for earthquake damage due to 7.6 and 8.6 magnitude events on the northern extreme of the NMSZ. One published document has covered the technical aspects of damage from seismic events for one major small city of southeast Missouri (Allen and Hoshall, 1986). No major published documents have covered the other small major southeast Missouri or northeast Arkansas cities. Because the largest, most active portion of the New Madrid Fault starts and ends in this region, they should receive support for seismic damage studies during the decade of the 90s.

## Seismic Hazard Awareness

Unawareness of seismic hazards is often common for those in regions that are located near known faults with a long period between occurrences. This imposes a particular hardship on waterworks operations in small cities located near the New Madrid Fault. It has proved difficult for money-strapped utility administrators to be concerned about events that occur once or twice every 100 years, let alone those that occur only once or twice every 500 years. This civic government's viewpoint, coupled with the regional Civil Engineers' general limited experience and knowledge of seismic risk data leads most of those engineers practicing in this region to assume that the greatest ground motion that a particular site will experience is that which is reflected in seismic building codes, (which apply to new seismically designed ductile buildings and structures). It is not commonly known by most regional engineers that building code calculated design accelerations are not true scientifically predicted ground accelerations. The seismic design maps and prescribed calculations in the seismic building codes for new buildings do not make this point clear. This is very confusing for new civil engineering entrants into the seismic arena, and it typifies the problems to be expected in any new technology start-up program.

## Historical Earthquakes in Arkansas

The large earthquake history of eastern Arkansas in the 19th century shows that on December 16, 1811 two earthquakes occurred on the New Madrid Fault in northeast Arkansas, one of magnitude 8.3 Ms and the other of magnitude 8.5 Ms (Stewart and Knox, 1991). Both resulted in near field high damage intensities that have been typically estimated as MMI X-XII (Algermissen and Hopper, 1984). These two earthquakes were followed by two equal or larger magnitude earthquakes in January and February of 1812 in Missouri. The last and largest was nearest to New Madrid, Missouri, resulting in this name, "New Madrid Earthquake," being given to all of the earthquakes. On January 4, 1843, the next major New Madrid Earthquake occurred on the southern end of the active portion of the fault near Marked Tree, Arkansas, with an approximate magnitude of 6.0 and an estimated maximum damage intensity level of MMI = VIII (Nuttli, 1990). These several historical earthquakes are the driving force behind earthquake preparedness in the central United States. The last such event greater than 6 was in 1895 at Charleston, Missouri, which is near New Madrid, Missouri, Ms = 6.7.

Arkansas Major Small City Waterworks at Risk

Earthquake experience in the western states indicates that near field effects of a magnitude 7+ earthquake can be very damaging to waterworks facilities at locations 50 km or less from the source, and moderately damaging to facilities 50 - 100 km from the source. In the near field all frequencies of waves can be present, thus causing more extensive damage to the typical aseismically designed masonry and concrete buildings and structures utilized as waterworks facilities in the Central United States.

In eastern Arkansas there are 8 major small cities located less than 100 km from the known active portions of the New Madrid fault, as shown in Figure 1. Associated MMI intensity ratings for a magnitude 7.6 earthquake (Nuttli, 1990), are shown in Table I.

**TABLE I - MAJOR SMALL ARKANSAS CITIES**

| City | County | Population | Fault Dist km | MMI |
|---|---|---|---|---|
| Blytheville | Mississippi | 23,844 | 20 | X |
| Forrest City | St. Francis | 13,803 | 60 | IX |
| Jonesboro | Craighead | 31,530 | 40 | VIII |
| Marianna | Lee | 6,220 | 90 | IX |
| Paragould | Green | 15,248 | 40 | VIII |
| Pocahontas | Randolph | 5,995 | 80 | VII |
| Walnut Ridge | Lawrence | 4,152 | 80 | VIII |
| West Memphis | Crittenden | 28,138 | 40 | X |

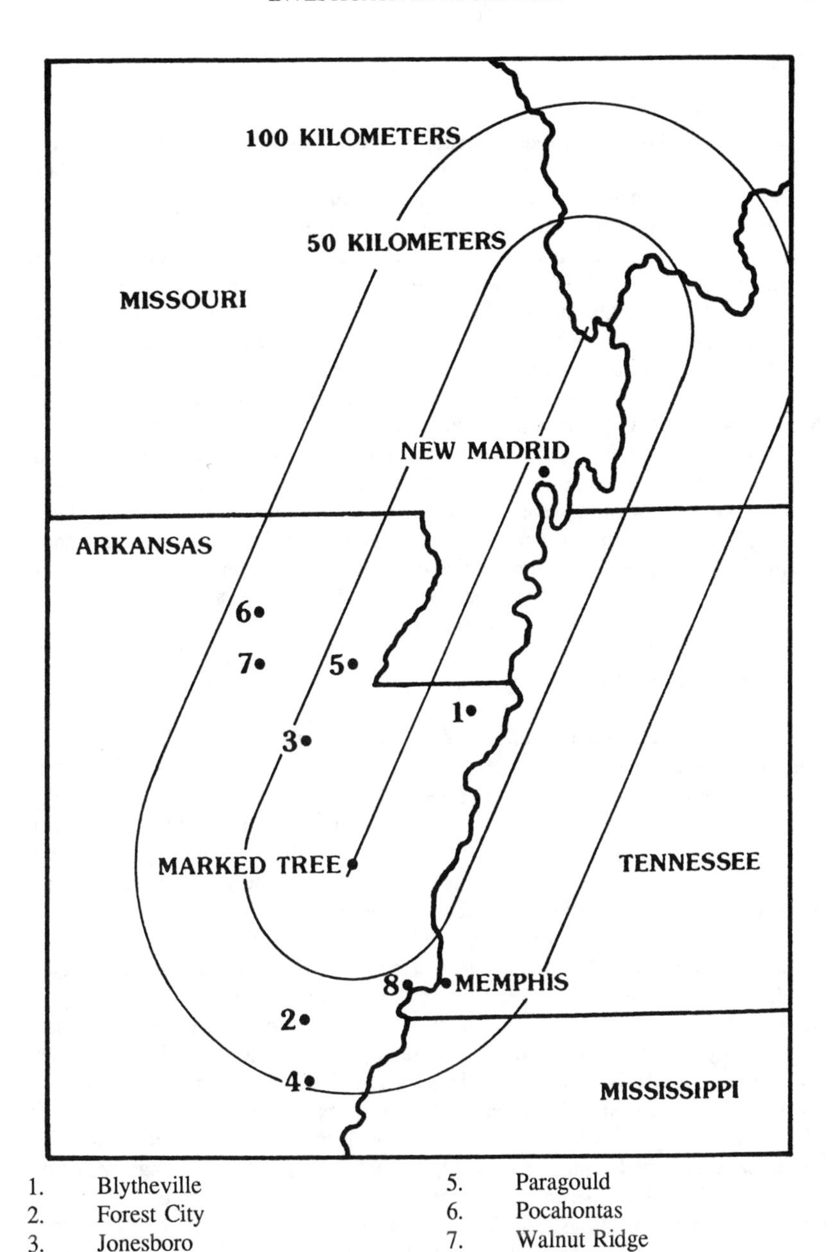


1. Blytheville
2. Forest City
3. Jonesboro
4. Marianna
5. Paragould
6. Pocahontas
7. Walnut Ridge
8. West Memphis

Figure 1 - Arkansas Cities Effected By the New Madrid Fault

## Discussion

### Lifeline Related Impact

Of the eight scenario earthquakes considered in a recent study (Scawthorn, Khater and Rojahn,1991), it was found that one of the greatest lifelines-related impacts to the United States economy would result from a magnitude 8+ event on the New Madrid fault. "Because much of the U.S. national infrastructure was constructed prior to modern seismic design considerations, the U.S. is faced with the problems of existing lifelines which are seismically vulnerable and which, if disrupted, could result in major economic losses, and probably environmental damage and human injury." The study concludes that electric power and waterworks are the two most critical lifelines. This study took the data developed on the basis of expert opinion in the ATC-13 project (Applied Technology Council, 1985), as a starting point for its lifeline vulnerability function and modified them per procedures outlined in ATC-13 for other states by shifting them one or two intensity units up for nonstandard construction (i.e., +1 or +2). Nonstandard refers to facilities that are more susceptible to earthquake damage than those of standard construction. "Older facilities designed prior to modern design code seismic requirements or those facilities designed after the introduction of modern code requirements, but without their benefit can be assumed to be nonstandard. Standard is assumed to represent existing California facilities ..." (Scawthorn, Khater, and Rojahn, 1991).

### Damage to Modern Nonstandard Buildings

The expected damage effects of the next major New Madrid earthquake on the buildings and structures in NMSZ, appears to follow the "non-standard" rule, although not without question. Very little data exists concerning the seismic damage history of well-built, well-designed, modern unreinforced masonry buildings. It is clear from an examination of damage reports from earthquakes of the past 50 years, that poorly-constructed, poorly-designed, and structurally deteriorated older unreinforced masonry buildings can be expected to perform poorly during a major earthquake, especially when built in areas underlain by soft, weak, unconsolidated or liquefiable soils. What is not clear or readily obvious is the damage level of moderate earthquakes to modern, well-built, well-configured, well-designed unreinforced masonry buildings. A close reading of major earthquake damage reports will show that seldom are these building types mentioned in the damage list. One reason for this, is that for several decades unreinforced buildings and structures have been discouraged or forbidden in western states, even those that are of the best design attributes. This information coupled with the knowledge that several new seismically designed reinforced concrete, wood and masonry buildings with poor design configuration have failed in several fairly recent major events, leads one to assume the worst for all aseismically designed buildings that might be subjected to seismic forces in the Central United States. This data has confused engineers concerning the safety of these aseismic buildings, often making it difficult to distinguish between those that will perform well and those that will not perform well.

This is the legacy of following closely the ongoing history of earthquakes in the western states without paying proper attention to the true seismic strength of the inventory of newer well-configured, well-designed, well-built aseismically designed building types. It will be important in future studies, to single out the worst of the modern aseismic building types and the best of these modern building types and arrange all others according to a varying scale somewhere in between.

Rapid Visual Inspection

By looking at such well distributed nationally-acclaimed documents as Rapid Visual Screening of Buildings for Potential Seismic Hazards, 1988, ATC-21, the extent of the problem can immediately be seen. For example, ATC-21 was utilized recently in a rapid visual building inspection to access the vulnerability of all principal buildings in a small Arkansas town. The resulting "Structural Score" for all buildings was well below the recommended Structural Score threshold for acceptable buildings types recommended in the ATC-21 report. Although this placement might lead to the correct conclusions for all of these buildings, a question arises concerning the database damage information utilized to develop the method. Did it include well-configured, well-designed, well-constructed modern unreinforced masonry buildings? For future waterworks building projects, it seems clear that no one should recommend the construction of new unreinforced masonry buildings in known hazardous seismic zones. The addition of steel reinforcement, diaphragms, and connections are relatively inexpensive and new building code laws will require that buildings have code-approved seismic strength. But for now, these buildings do exist and they must be properly evaluated as part of a complete lifeline vulnerability assessment for the cities in the New Madrid area.

Seismic Loss Estimates

The estimation of strong ground motion requires the use of strong-motion attenuation relationships and mathematical expressions relating strong-motion parameters, such as peak acceleration, to characterize parameters of the earthquake and site, such as earthquake magnitude, source-to-site distance, and site conditions. Attenuation relationships represent the phenomenon of a chosen parameter such as amplitude diminishing as seismic waves propagate from the earthquake source to the site (Taylor, 1991). Too little is currently known about this and other near field parameters in the NMSZ. The attenuation relationship should predict a strong-motion parameter that is appropriate for the intended application. The most frequently used strong-motion parameter relationships predict peak acceleration, peak velocity, and/or peak displacement.

Analysis of Hazard, Risk and Analytical Computation

"A ground shaking hazard analysis can be either deterministic or probablistic. In a traditional deterministic analysis, one or more earthquakes are hypothesized to occur near the site and estimates of ground motion are computed from an attenuation relationship." "In a probabilistic assessment of ground-shaking hazard, all potential earthquakes and their associated probabilities of occurrence are factored into the analysis." (Campbell, 1991). It would be good if a published study were readily available giving deterministic estimates of strong ground motion for major small cities in the NMSZ.

Probabilistic assessments of the ground shaking hazards in Arkansas as presented in the Standard Building Code give rise to middle of the road predictions of building safety. Since most existing Arkansas utility buildings and structures are of an aseismic design, i.e., non-ductile and perhaps ill-configured or poorly-sited for earthquake resistance, use of the more specific deterministic damage assessments to calculate probable accelerations from the ground motion of a specific earthquake scenario, i.e., using a specific worst-case earthquake source at a given magnitude, duration and a specific date, might be best. The central issue would be to consider adverse effects on a waterworks facility subjected only to an earthquake every 100 years rather than one that could also occur every 50 years or 200 years. A probabilistic assessment of the ground shaking hazard that shows a reduced value of ground acceleration, because of a longer recurrence period could cause waterworks facilities to suffer needless damage when the event finally occurs, if the design requirements have been reduced using probabilistic data.

In northeast Arkansas there has not been a major damaging earthquake in 150 years. According to most published recurrence relations, the 1843 Marked Tree, Arkansas, earthquake is past due (Algermissen, 1989 and Hamilton and Johnston, 1990). However, with this influence, most published earthquake design acceleration curves reflect less danger rather than more danger. This has resulted indirectly in a false feeling of security from smaller damage estimates based on those reduced probabilistic assessments.

## Risk Assessments Must be Balanced

"Confidence in modeling future earthquake sources is highly relevant to vulnerability reduction because response analysis of critical engineered structures requires the definition of an earthquake source that is well established by geological and seismological analysis." (Bolt, 1991). Probability models have also been used to prepare ground shaking hazard maps for the whole United States and for specific regions, for example, the NEHRP earthquake parameter maps attached to the 1988 Edition of the NEHRP Recommended Provisions for the Development of Seismic Regulations for New Buildings (FEMA, 1988). "These maps have now been incorporated in official building codes, with the explicit understanding that a balance of risk is implied between the odds of larger shaking and the high cost of over-design." (Bolt, 1991). The difficulty arises when the local civil engineer lacks this explicit understanding.

## Northwestern States Waterworks Studies

In a recently published earthquake loss estimation study for the risk evaluation of the Portland, Oregon Waterworks, a detailed methodology has been presented. "The loss estimation model provides a tool for city organizers, officials and emergency response personnel to identify the critical areas in need of pre-earthquake attention and to identify systems that need to be retrofitted." "Seismic loss estimation curves have been developed for various facilities in the ATC-13, entitled Earthquake Damage Data for California." (Elliott and Ballantyne, 1991). ATC-13 (Applied Technology Council, 1985), was utilized as the basis for loss estimation curves in the Portland study, resulting in the development of loss curves that provided a method of predicting how much damage a certain type of facility is expected to experience for a range of earthquake intensities in California. "The curves do not necessarily provide a good method for other regions of the United States." (Elliott and Ballantyne, 1991).

The seismic vulnerability assessment of the Seattle Water Department's water system facilities was presented in a 1991 ASCE lifelines publication (Anton, Polivka, and Harrington, 1991). "One of the important steps in conducting a seismic vulnerability assessment is to establish an appropriate seismic criteria to use in the evaluation." Response spectra were established for use in the Seattle Water Department (SWD) study with 0.5, 3, 5 and 7% damping; representing horizontal motion at the ground surface for a maximum credible earthquake, a 1 in 100 years return period earthquake and for 3 types of soil conditions. Based on this seismic criteria, a seismic assessment of the inventoried facilities of SWD facilities and pipelines for all classes of earthquakes considering both structural and nonstructural components and equipment was completed. The SWD assessment consisted of data reviews, site visits, and preliminary seismic evaluations.

## Suggested NMSZ Waterworks Seismic Criteria

The seismic criteria for NMSZ waterworks facilities must be based upon historical seismicity and experience of damaging earthquakes in other regions. Because of the range of earthquakes possible or probable on the NMSZ, it is proposed that the waterworks facilities in Arkansas, Missouri and the other central states potentially affected by future events be evaluated for at least three levels of earthquakes, magnitudes 6.5, 7.5 and 8.5. The appeal of this recommendation is that it not only includes most possible seriously damaging earthquakes in the Central United States, but also those of recent history in the western United States.

By using these three major magnitudes, it could make it easier for Civil Engineers to visualize, as well as calculate potentially damaging effects of future earthquakes on the New Madrid fault, by comparing them to similar-sized well-documented modern events in California.

## Exposure Time For Utilities

The proper exposure time for a waterworks facility is longer than the recommended exposure time for most other structures, because the water utility company must perpetually stay in operation to support the regional water needs, i.e., waterworks will not close unless everything else closes. Therefore, unless the facility is moved, it will need to be evaluated for a nominal time of 100 years or longer.

## The Potential Losses Considerations

The potential losses for a modern re-occurrence of a major New Madrid Earthquake would be large, especially for a worst case scenario, i.e., for a large earthquake of the size of one that has occurred in the past, to recur at the location and time of day that would produce the maximum destruction at a given site. "Analysis of the intensity patterns of the 1811-12 earthquakes and other earthquakes in the Central United States in conjunction with present-day distribution of dwellings gives an upper-bound loss estimate of about $50 billion to dwellings (in 1980 dollars) for a re-occurrence of the 1811-12 sequence" (Algermissen, 1989). "The expected maximum loss to dwellings in a 50-year period (at a 10 percent chance of being exceeded) is about $5.5 billion (1980 dollars), and a magnitude 6 to 7 earthquake would cause approximately $3.6 billion in dwelling losses." (Hamilton and Johnston, 1990).

Losses to dwellings could be increased or reduced due to the success or failure of municipal water systems to remain operable. It is difficult to calculate the risk for lifeline survival on potential loss to dwellings, but it is safe to predict that disruptions in water supply due to either disruptions in electric power or damage to concentrated water facilities, including pump stations, storage facilities, and treatment plants, or water delivery systems, would increase the maximum loss for any major small city.

## Risk Assessment Procedure

In any water system risk assessment conducted today in the NMSZ, it is necessary to start with a less than finished or accomplished procedure. This is true, even when trained outside engineering investigators are utilized. The principal reason is that the usual range of background information normally relied upon in some western states is not available in Arkansas, Missouri and the other central states. When considering the earthquake sources and magnitude for a specific site, the range of accuracy of variables selected must be considered, since there are no near source recordings from medium or large earthquakes in the NMSZ.

## Field Experiences and Failure Considerations

Recent seismic vulnerability assessments of actual lifeline and utility facilities in Arkansas and Missouri have revealed several types of major buildings, including poured concrete frame, tilt-up concrete panel, pre-cast concrete frame, concrete unit masonry and brick masonry. With few exceptions the buildings observed have all contained major structural deficiencies. The most serious deficiencies observed have been in buildings that had undergone several levels of periodic modifications for enlargement purposes. Bar-joist supported floors and roofs without either overlaid diaphragms or wall connections were noted to be very common. Prefabricated concrete floor and roof supports without overlaid structural diaphragms or wall connections were also found to be present.

Civil Engineers must develop the predicted response of each structure whose primary structural elements could be overloaded during ground shaking. Included in the assessment should be the catastrophic collapse potential due to failure of major structure elements caused by soil-rock/structure resonance. Configuration problems are very important when considering the possible modes of the structural failure. Things to look for include vertical discontinuities (soft stories and offsets), plan view irregularities (torsion, lack of symmetry, and re-entrant corners) and weaknesses in structural details.

## Usefulness of MMI Values for Vulnerability Assessments

The preliminary estimations of damage in the NMSZ to lifelines could be based on an estimate of Modified Mercalli Intensity (MMI). In the NMSZ, predicted intensity levels are a better descripter of ground motion for the general public than magnitude. They are convenient, easy to understand and widely available.

Equations relating Peak Horizontal Acceleration to MMI and site conditions have been developed, but even in California, these are recognized to have a larger standard error than parameters estimated from source magnitude and distance equations, which implies that there is only a weak physical relationship between intensity levels and peak ground motion parameters.

On the plus side the MMI values are particularly advantageous in Arkansas for preliminary work because they include the effects of the amplitude, frequency content and most importantly, the duration of expected ground motions as experienced in past events. MMI values could serve as a measure in the NMSZ for predicting earthquake damage levels, because of the qualitative relationship to those aspects of ground shaking that cause damage.

## Preliminary Structural Risk Evaluations in NMSZ

In performing a preliminary structural risk evaluation in Arkansas or Missouri, the engineer can utilize any one of the numerous USGS, NEHRP Provisions or code parameter maps to determine which sites should be rated the most hazardous. It is also possible to determine the approximate distances to the fault zone. This information, coupled with knowledge of the underlying soil/rock characteristics, can be studied to determine the likely ground motion signature at specific sites.

Local ground conditions have a significant effect on strong ground motion at a site. As a minimum, the difference between shallow and deep soils and between soft and hard rock are necessary to characterize site effects. Engineers must use judgement concerning the estimated effects of local site conditions on the published intensity values. USGS published maps in the FEMA Six Cities Report (Allen and Hoshall, 1983) gave MMI ranges in Little Rock, Arkansas, of VI to VIII, where as the average for the city is shown to be VII.

Information from several recent modern earthquake events shows that unstable foundation soils are those most likely to cause the greatest damage. The damage in San Francisco, Oakland, Santa Cruz and Moss Landing during the Loma Prieta earthquake clearly demonstrated this fact.

## Necessary Strength Requirements

Unfortunately, there is very little agreement at this time concerning the recommended level of strength necessary to protect the typical aseismically constructed lifeline building or structure from earthquake damage. Essentially, the task of determining recommended values is much larger than preliminary project lengths and support have allowed. Thence, the answer to the question of, "How strong is strong enough?", reduces to application of engineering judgement. Some engineers will prefer the conservative approach resulting in seemingly over-design, and others will choose to accept lesser values of strength closer to code related designs. Without additional research, there will be very little accuracy in these determinations.

## Steps In Risk Evaluation

A preliminary structural risk evaluation should address each item in the list of steps that follows:

- Step 1 Consult seismological data maps and evaluate potential earthquake hazard.
- Step 2 Consult geoscientific maps and evaluate existence of hazardous soil conditions.
- Step 3 Perform structural site inspection and determine the absence or existence of hazardous building configurations, materials, connections, foundations, walls, diaphragms or contents loading.
- Step 4 Prioritize perceived hazards at each site and all sites in a single project.
- Step 5 Prepare reporting documentation and make recommendations concerning needed detailed structural evaluations.

## Judgement of Preliminary Evaluation Report

At this point in the preliminary structural risk evaluation, it will be the responsibility of the engineer to choose the proper stance concerning the perceived site risks observed. Civil Engineers do not now have a standard one-plus-one equals two method of adding existing seismic hazards to obtain relative levels of risk, but clearly this is the direction of current earthquake engineering research and advanced practice. Lifeline risk analysis methods development has received special attention from the ASCE Technical Council on Lifeline Earthquake Engineering in its most recently sponsored technical publications (Eguchi, 1986; and Taylor, 1991). Future research of seismic events should reveal the relative risk associated with each category of structural deficiency, but until better information becomes available Civil Engineers will need to use a weighted combination of the seismic hazards at a single site to determine the facilities risk.

## Summary

Seismic vulnerability assessments of Central United States waterworks facilities will be performed by many of the region's Civil Engineers during the decade of the 90s. Practicing Civil Engineers need to become acquainted with this new technology. Only a very small percentage of those involved with the day-to-day operation of waterworks facilities in Arkansas, Missouri and the other affected states contiguous to the NMSZ have had experience and education related to earthquake engineering practice. For this reason, during the decade of the 80s when the central earthquake states were becoming more aware of their earthquake threat, most seismic vulnerability risk assessments were conducted either wholly or in-part by earthquake engineers from outside of the region. As this state-of-the-art in seismic engineering practice has evolved, too little attention has been focused on educating and training local engineers to perform these tasks. This is partly true because even on the west coast, many such vulnerability risk studies of waterworks facilities have been only recently completed. The considerable progress that has occurred during the past five years in this area, coupled with the ongoing engineering and seismological research and data development efforts makes the coming years of the 90s seem very promising for lifeline Civil Engineers who want to learn this new technology.

Two important issues have been emphasized in this paper. One, the shortage of Civil Engineers in the New Madrid states with seismic vulnerability risk assessment capabilities and two, the shortage of technical seismic data concerning attenuation, recurrence and magnitude of future earthquakes.

Recent federal, state and local activities to address these problems gives rise to the hope that future generations of Civil Engineers will be given adequate education and training to provide them with the knowledge, methodology and funding to help the citizens of these lesser known earthquake-prone areas to achieve a higher level of safety in their utility and lifeline systems.

## References

Algermissen, S.T., 1989, "Earthquake Hazards and Risk Evaluation in the Central United States," Overview of Earthquake Hazard Reduction in the Central United States, Central United States Earthquake Consortium (CUSEC), Memphis, Tennessee, pp. C1-C40.

Algermissen, S.T., and Hopper, Margaret, 1984, Estimated Maximum Regional Seismic Intensities Associated with an Ensemble of Great Earthquakes that Might Occur Along the New Madrid Seismic Zone, U.S. Geological Survey, Reston, Virginia, Map MF-1712, Reprinted 1986.

Allen and Hoshall, An Assessment of Damage and Casualties for Six Cities in the Central United States Resulting from Two Earthquakes, Ms=7.6, and Ms=8.6, in the New Madrid Seismic Zone: Central United States Earthquake Preparedness Project. Memphis, Tenn., Prepared for Federal Emergency Management Agency, n.d.

Anton, Walter F., Polivka, Ronald M., and Harrington, Laurel, "Seismic Vulnerability Assessment of Seattle Water Department's Water System Facilities", Lifeline Earthquake Engineering, Proceedings of the Third U.S. Conference, Monograph No. 4, August, 1991, American Society of Civil Engineers (ASCE), New York.

Applied Technology Council, 1985, "Earthquake Damage Evaluation Data for California", ATC-13, C. Rojahn, Principal Investigator, Redwood City, CA.

Applied Technology Council, Rapid Visual Screening of Buildings for Potential Seismic Hazards: A Handbook, ATC-21, Federal Emergency Management Agency, Earthquake Hazard Reduction Series 41, July 1988.

Bolt, B.A., 1991, "Balance of Risks and Benefits in Preparation for Earthquakes", Science, Volume No. 251, pp. 169-174.

Ballantyne, Donald K., and Heubach, William F., 1991, "Earthquake Loss Estimation for the City of Everett, Washington Lifelines", Kennedy/Jenks/Chilton, Federal Way, Washington, May 1991, pp. 3-1 thru 3-54.

Campbell, Kenneth W., "Analysis of Ground-Shaking Hazard and Risk for Lifeline Systems", Lifeline Earthquake Engineering, Proceedings of the Third U.S. Conference, August, 1991, ASCE, New York, pp. 581-590.

Eguchi, Ronald T., 1986, Lifeline Seismic Risk Analysis-Case Study, ASCE, New York.

Elliott, William K., and Ballantyne, Donald, "Earthquake Loss Estimation of the Portland, Oregon Water and Sewage Systems Evaluation of Concentrated Facilities", Lifeline Earthquake Engineering, Proceedings of the Third U.S. Conference, Monograph No. 4, August, 1991, ASCE, New York, pp. 500-509.

Federal Emergency Management Agency (FEMA), Estimated Future Earthquake Losses for St. Louis City and County Missouri, Earthquake Hazard Reduction Series 52, June 1990.

Federal Emergency Management Agency (FEMA), NEHRP Recommended Provisions for the Development of Seismic Regulations for New Buildings, Earthquake Hazard Reduction Series 17, October 1988.

Fuller, Myron L., The New Madrid Earthquake (A Scientific Factual Field Account), U.S. Geological Survey, Washington, 1912.

Hamilton, Robert, and Johnston, Arch, U.S. Geological Survey Circular 1066, Tecumseh's Prophecy: Preparing for the Next New Madrid Earthquake, Denver, Colorado, 1990.

Hwang, H.H.M., and Lee, C.S., 1990, Site-Specific Response Spectra for Memphis Sheahan Pumping Station, Nat'l Center for Earthquake Engineering Research, Technical Report NCEER-90-0007.

Hwang, Howard, "Seismic Hazards for Memphis Water Delivery Systems", Lifeline Earthquake Engineering, Proceedings of the Third U.S. Conference, August, 1991, ASCE, New York, pp. 510-519.

Nuttli, Otto W., The Effects of Earthquakes in the Central United States, 2nd Edition, Revisions by David Stewart, May, 1990.

O'Rourke, Michael, and Russell, Darrin, "Seismic Vulnerability of Memphis Water System Components", Lifeline Earthquake Engineering, Proceedings of the Third U.S. Conference, August, 1991, ASCE, New York, pp. 520-529.

Scawthorn, C., Khater, M., and Rojahn, C., "Lifeline Seismic Risk Analysis Impacts of a Catastrophic Earthquake on U.S. National Lifeline Systems", Lifeline Earthquake Engineering, Proceeding of the Third U.S. Conference, August, 1991, pp. 725-736.

Stewart, David, and Knox, Ray, Field Trip Guide to Representative Earthquake Features in the New Madrid Seismic Zone, Field trip dates March 15-17, 1991.

Taylor, C.E., 1991, "Seismic Loss Estimates for a Hypothetical Water System: A Demonstration Project", Technical Council on Lifeline Earthquake Engineering, Monograph No. 2, ASCE, New York.

**Seismic Mitigation of the Memphis Water System**
**Kevin M. Poe[1]**

Abstract

Memphis, Tennessee is located in zone 2 of the 1988 Standard Building Code, close to the New Madrid Seismic Zone. The probability of having a seismic event of significant magnitude is high. It has been estimated by Johnson (1985) that before the year 2035, the probability of a 6.3 Ms magnitude earthquake is 86 - 97%. Memphis Light, Gas and Water Division, a publicly owned utility, is taking precautions to prepare for such an event. This paper will briefly describe the Memphis water system, outline areas of vulnerability and present the action plan developed to prepare for an earthquake.

The Memphis Water System

All of the water used in Memphis is drawn from an artesian aquifer through 166 independent wells. These wells supply 10 treatment plants through a series of collecting mains sized from 12" through 48" in diameter. These treatment plants have a total pumping capacity of 413.5 MGD (million gallons per day) and treatment capacity of 215 MGD. The total underground storage capacity is 118 MG (million gallons). There are 15 elevated storage tanks with a total capacity of 4.7 million gallons.

There are 731,408 people served by Memphis Light, Gas and Water Division (MLGW) through 224,848 services.

MLGW owns and operates the facilities that provide water to the Shelby County area except those areas served by other municipalities. MLGW does sell water to some of these municipalities during peak loading.

---

[1]Water Systems Engineer, Engineering Services Area, Memphis Light, Gas and Water Division, 220 South Main, Memphis, Tn 38104

Most of the existing pipe in Memphis is made of either cast iron or ductile iron. Prior to 1959 the pipe joints were made using molten lead. After 1959 primarily mechanical joint or slip joint pipe was used. There is currently 14,018,375 feet of distribution pipe in the MLGW water system. Approximately 7,200,000 feet of that pipe is made of cast iron with lead joints. The downtown and midtown areas are predominately served by this older pipe. All pipe installed since the late 1970's has been ductile iron with slip joints.

Water is pumped from the Memphis sands by vertical turbine electric driven pumps. The water quality is high before treatment averaging 0.47 mg/L iron, 0.011 mg/L manganese, 0.13 mg/L fluoride and a PH of 6.4. It is delivered to an aerator through collecting mains. The water is aerated through a series of coke trays, and flows by gravity to the filter building where it receives a dosage of chlorine before filtration. Dual media filters are then employed to remove iron particles from the water. After filtration the water is again dosed with chlorine. Fluoride and phosphate are also added at this time. The water is stored in eleven (11) underground concrete reservoirs with a total capacity of 118 million gallons.

High service pumps then pump the water though primary feeder or transmission mains (30"-24") to secondary feeders (20"-16") and distribution mains (12"-6"). The distribution mains are generally well interconnected with few one way feeds.

Figure 1 shows a simple representation of the flow of water in the Memphis System.

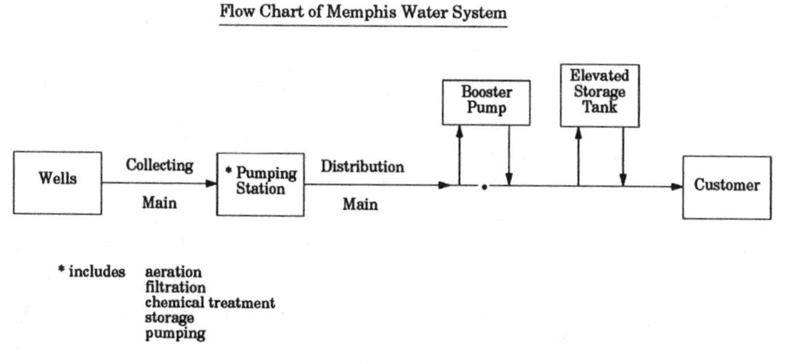


Figure 1

Areas of Vulnerability

Evaluating the Seismic vulnerability of the water system required analyzing the soil around Memphis, developing Seismic Hazard Curves, and determining what facilities were at risk.

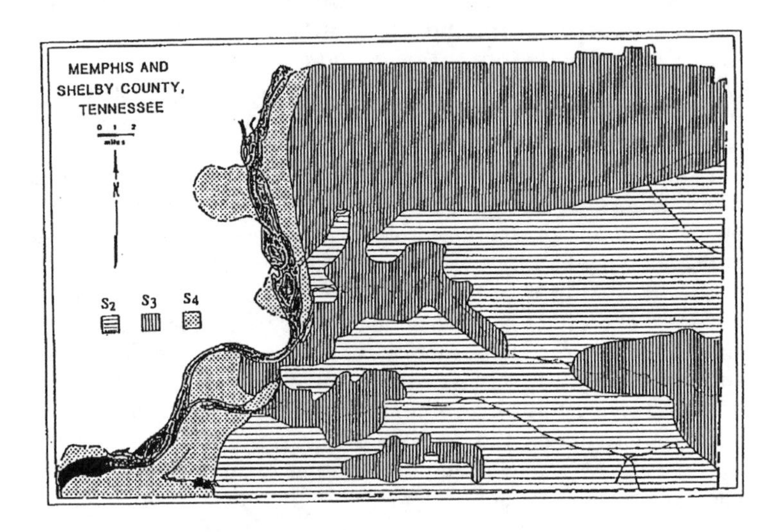


Figure 2. Soil Profile Classification Map - Site Coefficients $S_1$=1.0, $S_2$=1.5, $S_3$=1.8, $S_4$=2.0 (After Hwang 1991)

1) Soil Profile

The soil was evaluated by Hwang (1991). He classified the soil as $S_1$, $S_2$, $S_3$ or $S_4$ according to the Uniform Building Code. This classification was used to estimate the ground acceleration that could be expected in different parts of Shelby County for an earthquake of given magnitude. The information to develop this soil classification map came from soil borings in Shelby County (See Figure 2).

Since there is very little strong-motion data in the New Madrid region, a seismologically based model was used by Hwang to describe horizontal bedrock accelerations caused by shear waves generated from a seismic source. Hwang's study was based on simulating the effects of a New Madrid earthquake of moment magnitude M=7.5 located in Marked Tree, Arkansas.

This information was used by Hwang to develop a generalized map for liquefaction potential (see figure 3.) "The Liquefaction potential of a soil layer is affected by the relative density (sandy soils), percentage of clay (silty soils), effective confining pressure, grain-size distribution, and location of the water table. In addition the duration, frequency content, and amplitude of an earthquake also have significant effects on the liquefaction potential." Hwang (1991). The map shown in Figure 3 revealed only very limited potential for liquefaction in Shelby County for an earthquake of moment magnitude M= 7.5.

2) Seismic Hazard

The Seismic Hazard Curves developed to help evaluate the water system for weaknesses are shown in Figure 4. The curves were developed by Arch Johnson, Director of The Center for Earthquake Research and Information Center at Memphis State University. They show the annual probability of exceedence for different ground accelerations. There are three curves on the graph representing three (3) different parts of the County. The greatest seismic hazard is found in the northwest corner of the County which is sparsely populated at this time.

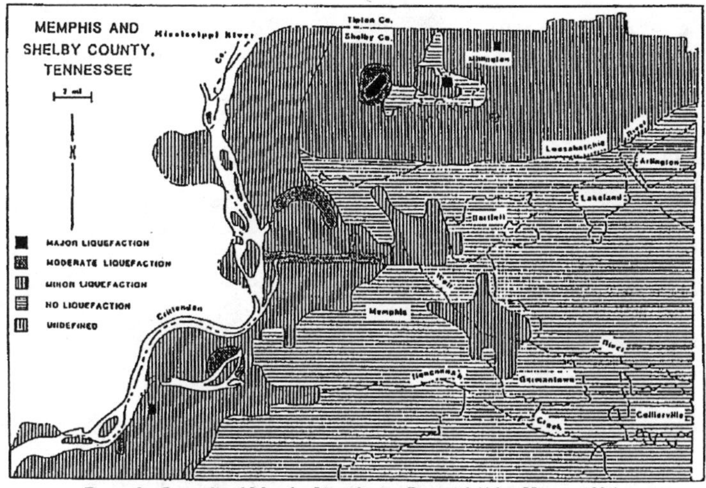


Figure 3. Generalized Map for Liquefaction Potential (After Hwang 1991)

Every critical facility was analyzed by Allen & Hoshall Inc. Engineers, Architects and Consultants under contract to MLGW. A value of seismic capacity was given to each facility.

Seismic capacity is the maximum ground acceleration an element is expected to sustain without damage and/or loss of service.

This seismic capacity was determined by structural analysis and observations from other historical earthquake damage.

3) Benefit:Cost Analysis

The following procedure was used by Allen and Hoshall to prioritize mitigation using benefit:cost ratios. "The determination of cost effectiveness of implementing mitigation of the seismic risks associated

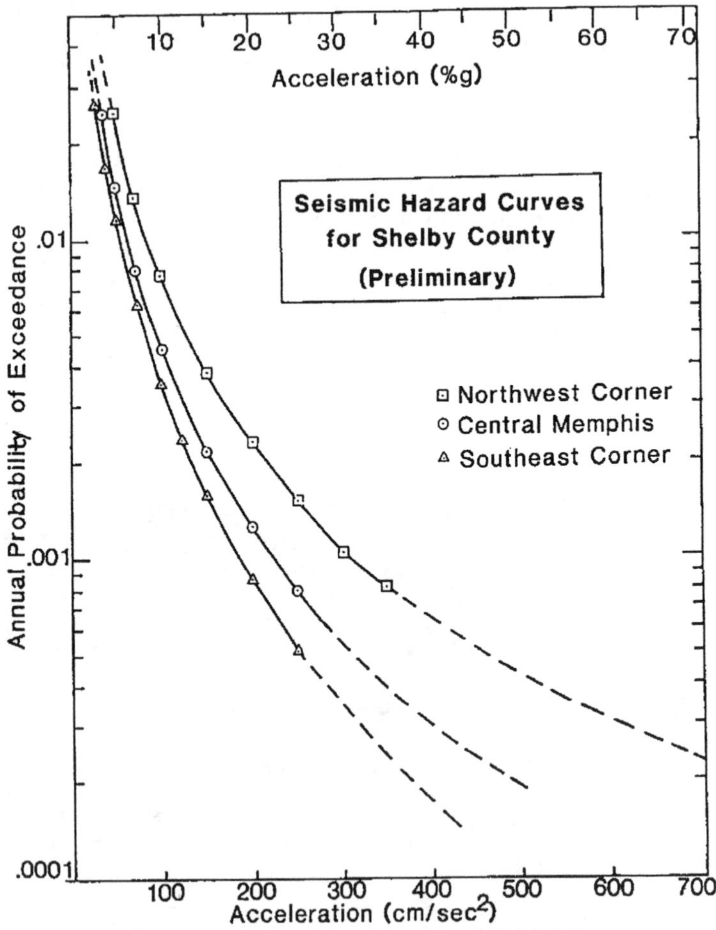


Figure 4. Seismic Hazard Curves (After Allen & Hoshall 1989)

with the identified deficient and marginal elements has been made by calculating a benefit:cost ratio (B:C) for each mitigation. The benefit of any particular mitigation is the value of the probable consequences avoided by means of the mitigation. Because of the significance of uninterrupted or promptly restored utility service with respect to a community's recovery from a damaging earthquake, the determination of the benefit associated with the mitigations considered here is more complex than for privately owned facilities. In the case of private facilities, the calculation of benefit could be made simply from the value of the facility to the owner and the cost of repairs. On the other hand, the benefit derived from mitigation of utility systems must include the value of the service to the utility customers as well as to the utility provider, MLGW in this case, along with the more easily determined repair or replacement costs. Cost, the other component required for calculation of B:C, is simply the estimated construction cost of any particular mitigative measure. The general procedure used in this study for calculating the B:C of any particular element is as follows:

1. The element is identified with respect to location in the study area and the appropriate one of the three Seismic Hazard Curves and the appropriate soil factor are assigned.
2. The ordinarily-expected service life and the current age of the element are determined and used to calculate the remaining life.
3. The existing capacity of the element with respect to seismic load is determined by structural engineering analysis.
4. Taking exposure period equal to remaining life, the results of steps 1,2, and 3 are applied to the Seismic Hazard Curves to determine the probability of exceedance of the element's existing seismic capacity during its remaining life.
5. The component of benefit representing the value of the element itself is established by estimating the cost of replacing the element.
6. The other component of benefit, the value of the element's service, is established by means discussed in detail on the next page after this step-by-step general presentation. It will be shown that, in most cases, the value of the service is the product of (1) the element's capacity as a proportion of the capacity of the system affected by its loss, (2) the anticipated length of interruption of the element's service and (3) the unit value of the service with respect to time.
7. The results of steps 5 and 6 are summed to yield the financial value of the loss of the element and its service.
8. The existing probable value of loss is calculated as the product of the probability of exceedance with respect to existing capacity (step 4) and the total value of the loss (step 7).
9. The capacity of the element with mitigative measures in place is determined by structural analysis. This is parallel to step 3, above.
10. The mitigated capacity of step 9 is combined with the information from steps 1 and 2 to produce the mitigated probability of exceedance, parallel to step 4.
11. The mitigated probable value of loss is calculated as the product of mitigated probability of exceedance (step 10) and the total value of the loss (step 7).
12. The benefit associated with the mitigation is calculated as the difference between existing probable value of loss (step 8) and mitigated probable value of loss (step 11). This difference is the value of probable consequences avoided.

13. The cost associated with the mitigation is determined by estimating the construction cost of implementing the mitigation.
14. Finally, the B:C of the particular mitigative measure is calculated as the quotient of the benefit (step 12) divided by the cost (step 13)."

Allen and Hoshall (1989).

In step 6 it was observed that the benefit associated with the mitigation of any element includes the value of that element's service.

The following discussion will illustrate how the value of service was calculated.

It was projected from Tennessee Statistical Abstract 1989 that the 1989 GCP (Gross County Product) for Shelby County was $15.2 billion dollars.

Earthquake damage to construction of all types will produce a reduction of demand for water service in the post-earthquake period of recovery. It was determined that in the post-earthquake situation, total demand will be about 72% of normal. This factor was applied directly to GCP, yielding $10.9 billion dollars.

The economic dependence on water service was estimated by percentage for the various components of manufacturing construction mining, trade and so forth. These were totaled and it was determined that 68% of the economy is dependent on water service. Application of the approximate 68% dependency to the GCP adjusted for building damage yields $7.4 billion as the annualized value of water service to the economy of the service area during recovery from a damaging earthquake.

The annualized value of water service was converted to a daily value of $20,300,000. The total treatment capacity of the MLGW system is 215 million gallons per day, yielding a value of $94,400 per million gallons. Allen and Hoshall (1989).

This information was used to determine where the money budgeted for mitigation could be most effectively spent. The following facilities have been evaluated for seismic vulnerability and prioritized for mitigation.

1. Water Pipes (mains):

Pipe lines, will most likely be damaged in regions of unstable soil, where sliding, liquefaction and ground rupture occur. The potential for landslides is low in Memphis since the area is relatively flat. Liquefaction is of minor concern in most areas (see figure 3).

As mentioned previously many of the water pipelines (collecting and distribution mains) in Memphis are made of cast iron with poured lead joints. These lines are the weakest link in the distribution system. Ductile pipe with flexible joints perform better than cast iron pipe with lead joints.

The cost of replacement would be exorbitant if done on a system wide basis strictly for seismic mitigation. These older cast iron pipes with lead joints will be replaced with newer more flexible pipes wherever possible.

All new installations are made using ductile iron pipe with slip joints. This pipe will allow $5^0$ deflection in each joint which translates to 18" per 20 foot length.

During the Loma Prieta earthquake, Pickett et al. (1991) documented that flexible joints that allow some rotation were effective. The major problem with slip joint pipe would be in translation. If there is a great deal of horizontal ground motion, the joints could come apart or telescope.

The old lead joint pipe is considered vulnerable but total replacement is not feasible. The newer ductile iron pipe with slip joints is not considered vulnerable.

Repair clamps for each size of pipe will be stocked. The graph shown in Figure 5 will be used to help determine the number of repair clamps needed for a Modified Mercalli Intensity of VIII.

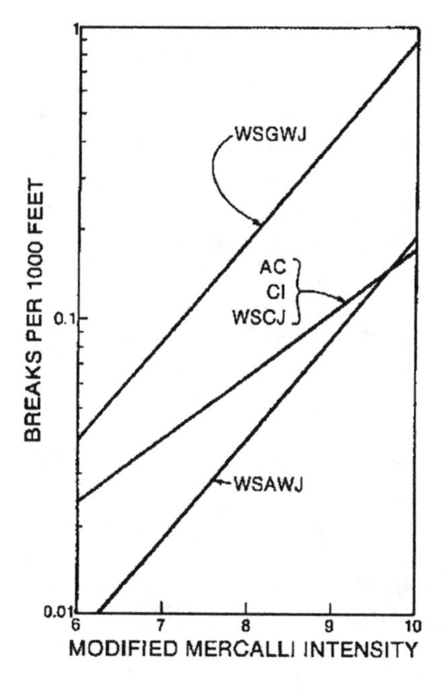


WSGWS = Welded Steel Gas Welded Joints
AC = Asbestos - Cement
CI = Cast Iron
WSCJ = Welded Steel with Caulked joints
WSAWJ = Welded Steel with Arc Welded Joints

Figure 5
Seismic Wave Propagation Damage for Underground Pipe (After Eguchi et al. 1983)

2. Wells

Memphis wells are approximately 300' to 400' deep with 20" steel casing surrounded by cement grout. Located inside the casing is a pump column pipe with vertical turbine pump bowls.

The wells could be subject to shearing forces on the casing, and piping connections, but should survive a 6.3 magnitude earthquake. Wang (1990) revealed that wells survived the 1987 Whittier Narrows Earthquake (Richter magnitude $M_L$=5.9) with little damage but there were outages due to loss of electric power. Since Memphis is relatively flat there is little danger from landslides. Liquifaction is of minor concern in the locations of the well fields (see figure 3).

The Memphis wells were generally considered to be resilient to ground accelerations estimated in this study. The pipe connections were considered vulnerable and are included in the mitigation.

3. Treatment Plants

The two oldest treatment plants Mallory (built 1923) and Sheahan (built 1933) were considered least resilient to ground movement. Each plant was evaluated and assigned a seismic capacity of 0.18g by Allen & Hoshall. The other plants were evaluated and assigned capacities from 0.22 to 0.45 g. A useful life was estimated for each plant. The Seismic Hazard Curve was used to determine probability of non-exceedance. The financial analyses of treatment plant mitigation, deal with alternative rather than additive courses of action. The installation of by-pass lines, emergency generators and pumps on the reservior for purpose of earthquake recovery protects the water system as well as structural reinforcement of the treatment plant buildings and anchorage of all equipment.

By-passes, emergency generators and pumps at the older treatment plants received the highest benefit:cost ratios.

4. Booster Pumps and Elevated Storage Tanks

The Booster Pumps and Elevated Storage Tanks received the lowest benefit:cost ratios. There are benefits to retrofitting tanks and booster pumps with flexible connections but the benefits are not nearly as great as those calculated for the treatment plants.

Elevated water storage tanks in Memphis, supply a relatively small percentage of the total storage capacity. As previously stated, the total storage is 123 million gallons. The elevated storage is 4.7 million gallons which amounts to 4% of the total storage.

Some of these tanks were installed several years ago in areas where the water distribution system was weak. The system has grown and improved since then so that these tanks are now of minor importance.

There are some tanks in the far reaches of the county that are still important but generally serve a small percentage of the service area.

It was estimated after structural analysis, that the tanks could resist 0.24g ground acceleration and the booster pumps 0.45g.

The last elevated storage tank installed was seismically designed according to American Water Works Association (AWWA) standard D100. The site was also evaluated to determine seismic risk. The Southeast portion of the County where this tank is located, has the lowest risk calculated for Shelby County.

Any new tanks or booster pumps will be designed with seismic risk in mind.

The booster pumps and storage tanks are vulnerable but did not receive a high priority in mitigation.

## Action Plan

An action plan was developed to implement mitigation measures identified for the existing system. A disaster response plan was also developed to address organizational needs during a crisis.

The following facilities were included in seismic mitigation.

1. Pipe
   Cast iron lead joint pipe will be replaced with ductile iron slip joint where it is economically feasible, as for example during some street improvements, system upgrading, etc.

2. Wells
   Key wells at each pumping station will be retrofitted for flexible connections to the collecting main. (See Figure 7.) The well "A" is rigidly connected to vertical pipe "B" through reinforced concrete member "C". "A", "B", "C" move as one unit. Flexible pipe fitting "D" is then connected to discharge pipe "E", allowing for movement in pipe "E" without damaging rigid connections to the well. There is no plan to rework the well casing or pump. Emergency generators will be provided at each plant to supply electricity to the pump motors.

3. Treatment Plant
   At treatment plants, by-passes will be installed so that water may be routed around the plant to enter the distribution system. This will require slight relaxation of water quality standards but is justified in light of the emergency. As mentioned previously

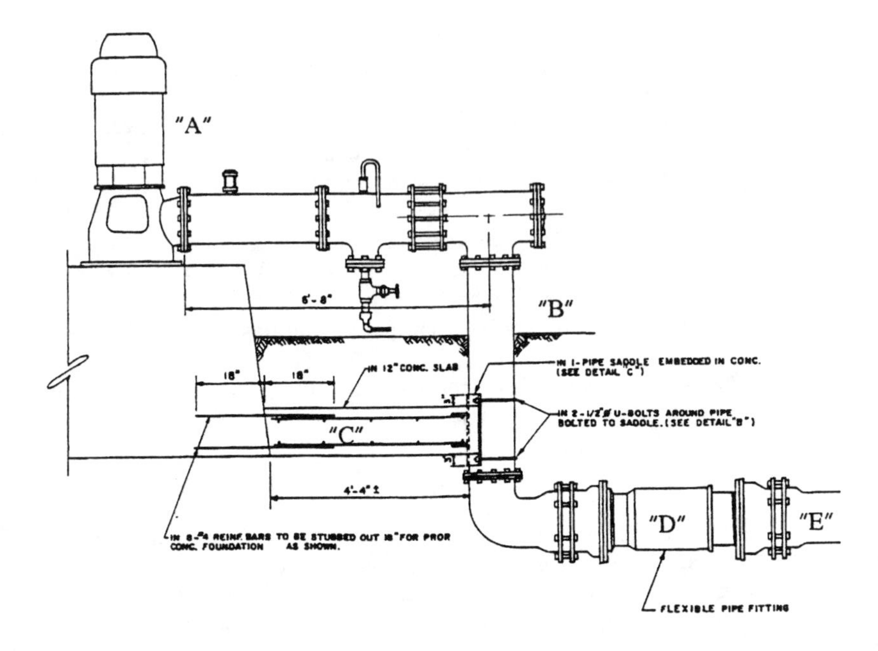


Figure 7. Seismic Retrofit for Wells

the water pumped from the artesian aquifer is potable without any treatment. Filtration and aeration do improve the quality by removing excess iron and carbon dioxide but the water is of excellent quality in its natural state. Emergency generators will be purchased for each plant and will have adequate electric capacity to run the necessary wells and high service pumps to deliver 15 MGD to the system. All battery racks at pumping stations will be secured as well as power transformers and switchgear.

The by-pass concept should provide adequate water for fire fighting and potable use even if the treatment plants are taken off line. These by-passes would only be used in the event of an emergency such as an earthquake. Periodically the generators and pumps will be tested and kept operational.

## Disaster Response Plan

A disaster response plan has been developed for the water system. Its purpose is to ensure an effective, professional and well organized response to a natural disaster or major water system incident. The following areas were outlined in the plan:

1. Activation guidelines for the plan.
2. Organization and responsibilities of a crisis management team.
3. Establishment of a Crisis Center and resources needed.
4. Employee responsibilities during the disaster response.
5. Philosophy for response, emergency repair, decontamination and restoration.

The Crisis Center will be headquartered in the Operations Facility. This is a new building designed according to seismic design criteria.

The Disaster Plan shall be activated by any of the following conditions:

1. A national disaster or major water system incident that imperils the integrity of the water system.
2. The water system is threatened by ground movement or undermining.
3. Any other circumstances where communications and control of an incident would be improved by using the Disaster Plan.

Critical materials needed to make repairs to the water system were identified and a list of items was developed. Larger fittings, 14" through 48" that are not used regularly were identified in the system as emergency items. The purchasing order point was set higher than normal, and a quantity of the materials was stored in a separate location. The graph in Figure 5 will be used to determine the quantity of repair clamps needed. This insures that the storeroom is not depleted of these critical items. Some of the items identified as "critical" were large pipe repair clamps 14"-24", mechanical joint sleeves, retainer glands, large diameter pipe and mechanical joint plugs.

## Summary

Memphis Light, Gas and Water Division has taken the initiative to prepare for a seismic event of magnitude 6.3 $M_S$. Experts in the field were employed to perform a Seismic Risk Assessment for MLGW. As a part of that study the first seismic hazard curve for Shelby County was developed and a soil amplification study was performed based on estimated peak horizontal bedrock accelerations.

The recommendations from these studies and others conducted on previous earthquakes have been used to develop an action plan to implement seismic mitigation to the Memphis water system.

This mitigation is reasonable considering the seismic risk involved.

Finally, much emphasis has been placed on seismic mitigation by Memphis Light, Gas and Water Division. Hopefully in the event of an earthquake, critical lifelines will remain operable for the citizens of Memphis, Tennessee due in part to the preparations stated herein.

Acknowledgements

The author would like to thank the following individuals for supplying information essential to writing this paper - Henry Winter, Tommy Whitlow, Junior Foreman, and Mark McAllister.

The data entry was performed by Ms. Carolyn Abston. Ray A. Ward and John Jackson aided in developing the graphics.

References:

Allen & Hoshall Inc. (1989) "Emergency Prepardness Study", Seismic Risk Assessment Study and Seismic Mitigation Plan MLGW Contract 9706. Volume 1 & 2.

Eguchi, R. T. Taylor, C. E., Hasselman, T. K., (1983) "Seismic Component Vulnerability Models for Lifeline Risk Analysis", Tech Rept. No 82-1396-2c, J. H. Wiggins Co., Rodonda Beach, CA.

Hwang, H. (1991) "Seismic Hazards for Memphis Water Delivery System", Proc. 3rd U.S. Conference on Lifeline Earthquake Engineering, Los Angeles, CA.

Johnson, Arch C. and Susan J. Nava, Journal of Geophysical Research, Vol.90, No. B7, "Recurrance Rates and Probability Estimates for the New Madrid Seismic Zone", July 10, 1985.

O'Rourke, M. (1988), "Mitigation of Seismic Effects on Water Systems", TCLEE/ASCE, National Convention.

Pickett, Mark, Laverty, Gordon, Abu-Yasein, Omar & Lay C. W., "The Loma Prieta Earthquake; Lessons Learned for Water Utilities.", Journal AWWA Volume 83, November 1991.

Wang, Leon Ru-Liang (1990), "A New Look into the Performance of Water Pipeline Systems From 1987 Whittier Narrows, Calif. Earthquake." Technical Report No. ODU LEE-05 In Life Line Earthquake Engineering Research Series, pp.27.

# Risk Based Decision Support Model for Water Delivery Systems Subject to Natural Hazards

M.A. Cassaro[1], Member, ASCE
M.J. Cassaro[2], R.K. Ragade[1], S. Alexander[1]

## Abstract

This paper presents a prototype model for analysis of potential damage and risk to water company systems in regions subject to earthquakes. Other hazards that produce loss to water distribution systems include: cold temperatures, floods, high wind, and tornado. Damage and loss produced by these events are incorporated into a risk model that considers repair and replacement cost, consideration of down time and loss due to reduced flow below demand levels. The model incorporates GIS, expert systems, and knowledge base tools to create a risk management tool that is capable of being updated with recent natural hazard records while maintaining current information on water system delivery facilities. A component of the model permits analysis of system effects for emergency response following a hazard event. Input of field observation of system damage during a natural hazard event permits flow analysis and supports decisions for setting priorities for repair.

## Introduction

Water delivery systems in the central United States are subject to the effects of natural hazards including earthquake. The effects of natural hazards usually are not considered in typical design practice, mitigation planning, repair and replacement, or the establishment of long-range strategy and policy to improve system reliability. The risk analysis model reported here is devel-

---

[1]Professor, [2]Graduate Student, Speed Scientific School, University of Louisville, Louisville, Ky, 40292

oped as a prototype for the Louisville Water Company (LWC), water distribution system. The model will assist water distribution managers to achieve acceptable performance under multiple natural hazard conditions. It will provide the opportunity to determine, with greater certainty, which components of the system are subject to breakdown under a natural hazard event, and to minimize interruptions of service.

Natural hazards produce wear and breakdown throughout the delivery system and affect components of the system in different ways. These effects must be considered within the routine planning and maintenance policies applied to the system. Although design practice is applied to these occurrences in some cases, integrated routine methods are needed that consider relative expectations of the loss that potentially damaging natural hazard events can cause.

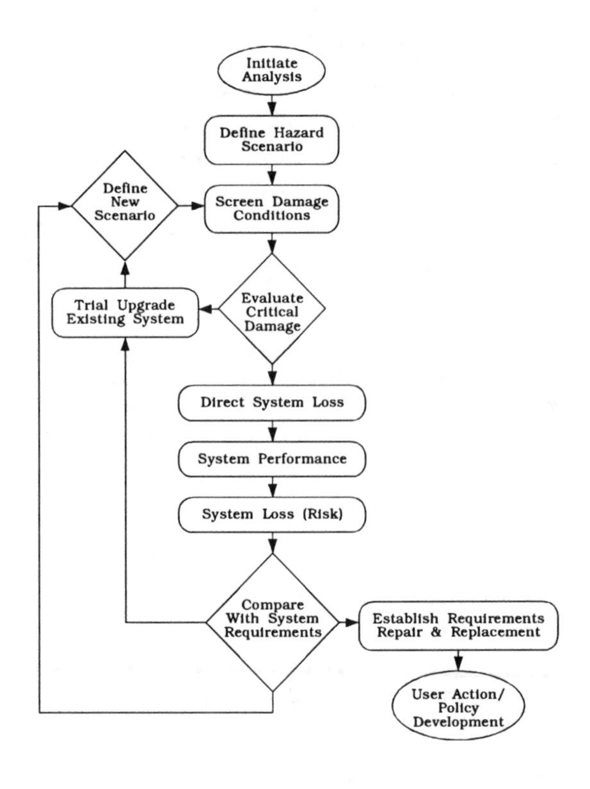


Figure 1 - Natural Hazards Risk Analysis of Water Delivery Systems - User Model

The model illustrated in Figure 1 permits the routine use of risk analysis procedures for policy development by water delivery decision-makers to provide information for applying repair and replacement strategy. The routine use of the model for planning will improve durability of the water delivery infrastructure and also the level of performance during hazards events.

Assisting Management Planning

The Risk Management Model in Figure 1 permits the user to define a hazard scenario for which the system risk is analyzed. Damage functions for each component in the system inventory are contained in a database that provides damage levels as a function of hazard effect as defined in Table I. By selecting a hazard and defining its magnitude, a hazard scenario damage analysis is performed from which the user may screen expected damage conditions and assess critical component survival requirements to maintain system performance. This evaluation permits the user to reject the disruption of critical components under the specified hazard scenario.

Those critically damaged components that must survive the prescribed scenario event are assigned a trial design upgrade in the existing system by simply rejecting the damage state predicted. If a new scenario is not selected, the damage conditions will now exclude disruption of critical components, and the analysis will consider these components to be functional under the chosen scenario. The critical components are then automatically listed in a file contained in the Repair and Replacement Action/ Policy Development Database, Figure 1, indicating that design capacity must exceed the level of hazard defined by the hazard scenario. This file also contains information about the percent damage expected and corresponding current design information for the existing component. This file can be used to achieve planned upgrade of the water system to improve reliability of the component.

The direct system loss is computed by combining each damaged component in the system damage data file with the relative replacement value record (Army Corps of Engineers, Hansen).

System performance is evaluated using KYPIPE (Wood) flow simulation software. The network flow model is modified to consider reduced nodal pressure and water supply due to the occurrence of breaks and leaks in the expected damage state. A series of simulations are performed as the modified network flow model is adjusted to compensate for negative pressures in the flow simulation results (Khater, Okumura). The resulting pressure and water flow are compared with the demand requirements at each node in the system. The results of this analysis permit an evaluation of system performance under the hazard scenario. Critical situations are evaluated such as satisfying pressure requirements to support firefighting demand.

The outcome of the loss analysis is used to provide design planning information to improve system reliability under specified hazard scenarios. Risk related to the probability of occurrence is presented in the form of functions (Taylor) relating probability of exceedance versus dollar loss for an applicable range of each hazard type. The results provide system planners with an ability to control cost while achieving maximum benefit.

The information to be provided to managers requires timely and reliable data about hazards and expected damage levels for each potential hazard scenario. Consideration of expected hazards and associated damage levels will permit control of design while providing for policy development based on reliable levels for accepting risk.

Items included in the processes of the risk management prototype program are:

- Define potential hazards impacting the water system as listed in Table I.
- Inventory the water delivery system facilities in a database as illustrated in Table II.
- Establish spacial reasoning capability with a Geographic Information System (GIS) for the water system facilities as illustrated in Figure 6.
- Create component damage functions by defining water delivery system component vulnerabilities based on specific hazard levels. An example is shown in Table II.
- Create loss algorithms for direct loss based on expected damage resulting from specific hazard levels (Taylor, Ballantyne, ATC, Cassaro, Burke).
- Determine probable system performance for specified hazard scenario.

Hazard Conditions

The data for hazards are dependent upon potential impacts to, and past experience of, water delivery systems and on their expected performance with respect to the hazard. Data providing earthquake hazard effects on water delivery systems is acquired from Taylor, ATC, Wang, and Trautmann. Other natural hazard data are derived from information obtained from the National Weather Service, Severe Storms Forecast Center, and National Flood Insurance Program. A knowledge base forming an expert system incorporates relevant knowledge to yield appropriate information for each natural hazard as given in Table I. Each hazard is defined by three factors:

1. Magnitude - This record establishes the intensity level of the hazard in standard terms.
2. Recurrence interval - This record provides the probable return period for each level of hazard magnitude.
3. Environmental effect - This record converts the magnitude of the hazard into terms that associate the hazard with damage effects to the system components.

Table I
Hazard Definitions

| Hazard | Magnitude | Environmental Effect |
|---|---|---|
| Earthquake (Figure 2) | Richter | Ground motion spectra, peak acceleration, MMI (Nuttli, Johnston, Hart), liquefaction (Esmaeelzadeh, Liao, Seed 1971, Seed 1984, Youd) |
| Weather (Figure 3) | Temperature | Monthly average minimum daily temperature |
| Tornado (Figure 4, 5) | Wind Speed | Pounds per square foot based on structure geometry |
| Flood | Flood Stage | Feet of water above ground level |

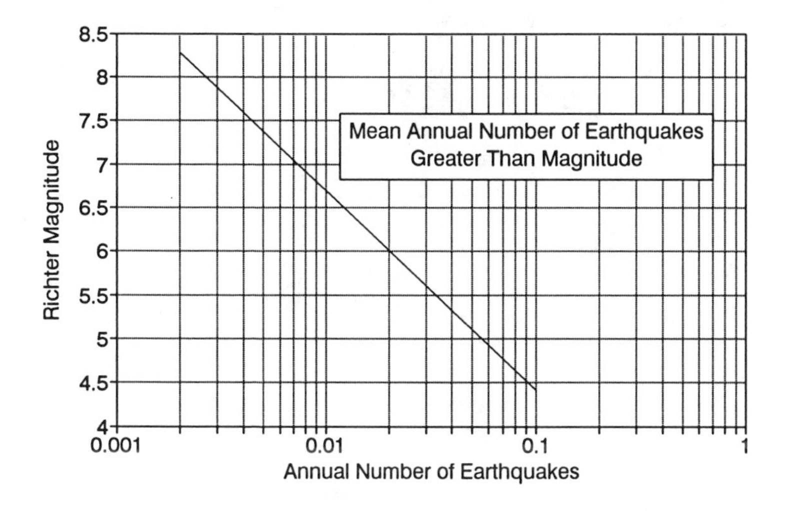


Figure 2 - Earthquake Hazard for New Madrid Fault

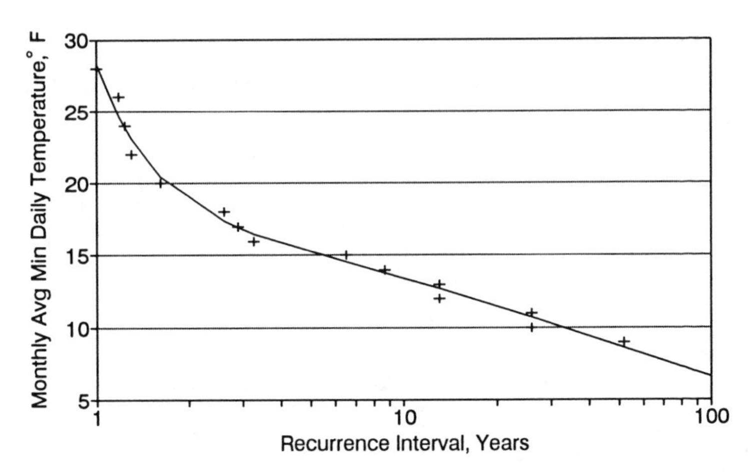


Figure 3 - Temperature Hazard for Louisville and Jefferson County, Kentucky

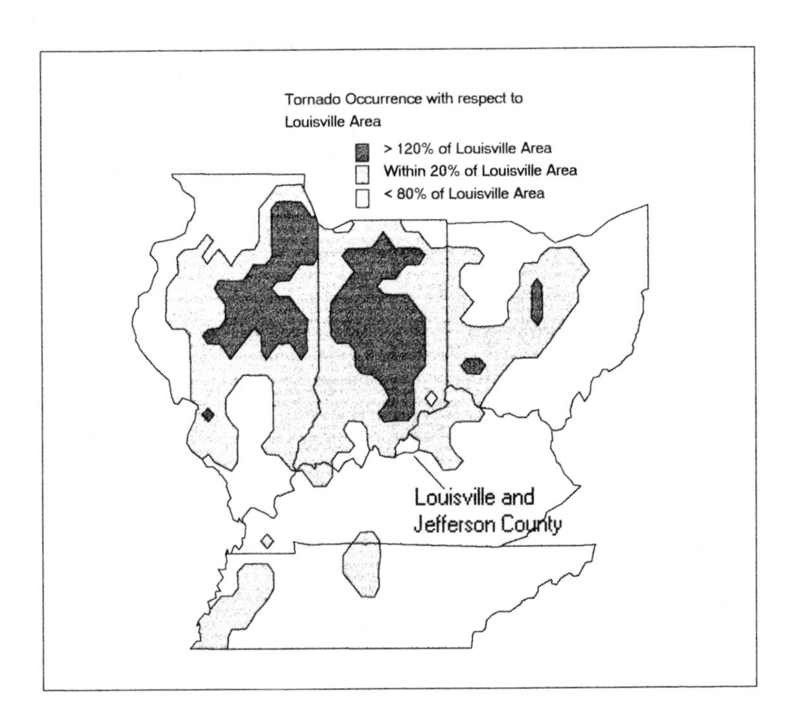


Figure 4 - Tornado Occurrences in the Five State Region Surrounding Louisville and Jefferson County, Kentucky.

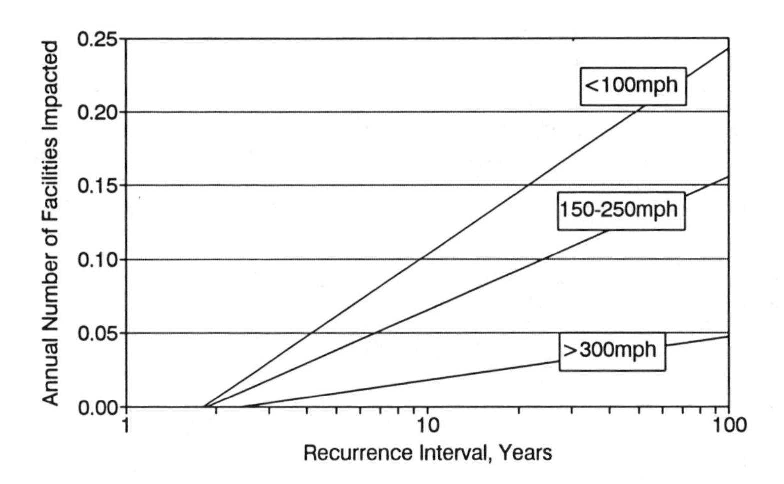


Figure 5 - Tornado Hazard for Facilities Owned By the Louisville Water Company

Inventory

The data included in the water facility inventory are: facility class, location, type construction, replacement cost and capacity. These data are contained within interactive knowledge base tables as illustrated in Table II. The knowledge base is capable of accepting changes to the water facilities based on repair, replacement, and new constructions. The information contained in these tables can be displayed spatially on GIS as shown in Figure 6. Some typical forms of data are:

- Pipe network files describing location, topography, pipe data, and water demand.
- Fixed facility files containing location, structure use classification, and structural data.

Vulnerability Analysis & Loss Estimation

The system vulnerability is determined from tables derived from component damage functions. For earthquake, component damage functions are based on a combination of categorization loss methods (ATC) and analytical techniques. The analytical technique applies structural dynamic analysis to each structure of the system, excluding buried pipes. The procedure applies determination of the local ground acceleration spectra computed from WAVES (Hart) based on expected bedrock accelerations (Nuttli) for a New Madrid earthquake scenario defined by Richter magnitude. The existence of an aquifer beneath Louisville, combined with a deep soil layer, raises the potential for large soil amplification of the bedrock accelerations shown in Figure 6. Analysis is performed by PAL-2 for storage tanks, ETABS and BARON (Baron) for buildings, and HAMBURG (An) for pump stations or tanks coupled with pipes. Damage functions for high wind and tornado are determined using similar analytical techniques under the action of static wind forces.

For a hazard scenario, the damage state of each fixed facility, $DS_f$, is determined probabilistically from the damage functions, and used in equation (1). Damage information is shown in Table II.

TABLE II
LWC Inventory and Damage Functions

| | Facility | | Damage Information | | | | |
|---|---|---|---|---|---|---|---|
| Location | Class | Facility | Source | Hazard | Intensity | % | Probability |
| B.E. Payne | 4 | Low Lift | . | . | . | . | . |
| | 3 | Filter Building | . | . | . | . | . |
| | 12 | High Lift | . | . | . | . | . |
| Crescent Hill Pump Station | 84 | Boiler | . | . | . | . | . |
| | 84 | Pump House | . | . | . | . | . |
| | 84 | Pump Annex | . | . | . | . | . |
| Crescent Hill Filter Plant | 84 | Chemical House | . | . | . | . | . |
| | 84 | North South Filter | . | . | . | . | . |
| | 84 | East Filter | . | . | . | . | . |
| | 84 | New East Filter | . | . | . | . | . |
| Zorn Station Pump Station | 84 | Pump Station 2 | . | . | . | . | . |
| | 84 | Pump Station 3 | Expert | Earthquake | MMI VI | 0 | 3.7 |
| | | | | | | 0.5 | 68.5 |
| | | | | | | 5 | 27.8 |
| | | | | | MMI VIII | 0.5 | 1.6 |
| | | | | | | 5 | 94.9 |
| | | | | | | 20 | 3.5 |
| | | | | | MMI X | 5 | 11.5 |
| | | | | | | 20 | 76 |
| | | | | | | 45 | 12.5 |
| | | | | | | 20 | 37.6 |
| | | | Expert | Tornado | 150mph | 5 | 20 |
| | | | | | 200mph | 20 | 45 |
| | | | | | 250mph | 25 | 35 |

Pipeline damage is determined using suggested damage functions (ATC, Taylor) for earthquake and Figure 7 for cold temperature hazards. The temperature damage function provides the number of breaks per mile per unit time. For example, Figure 7 illustrates the function for a one month time unit. The method of analysis displayed in Table III is used to evaluate the distribution of breaks by pipe material.

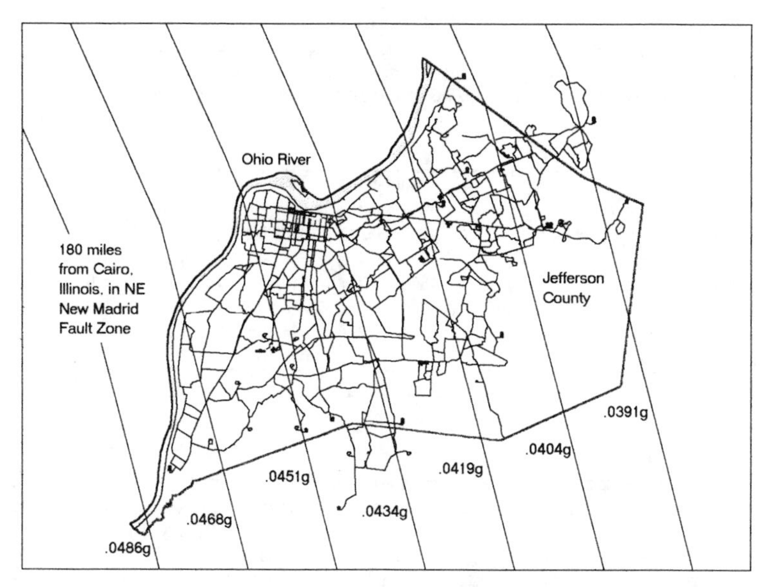


Figure 6 - Louisville Water Company Pipe Network and Fixed Facilities with Bedrock Accelerations for a 7.5 Richter Magnitude Earthquake at NE New Madrid Fault.

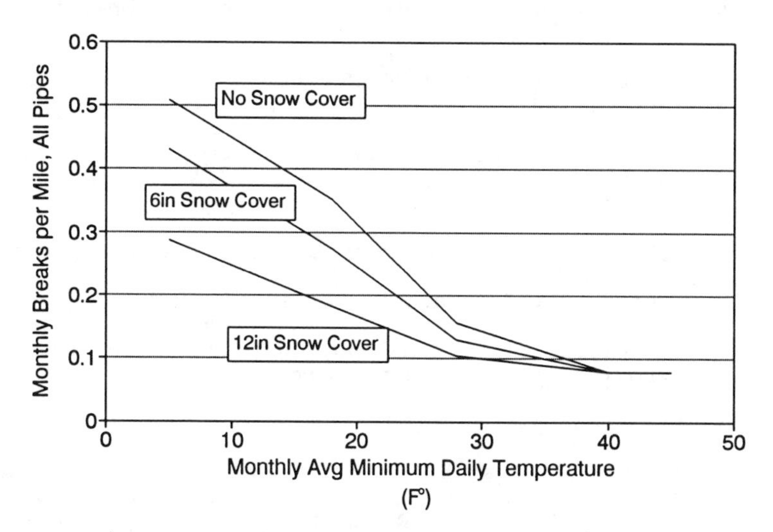


Figure 7 - Temperature Damage Function for Buried Pipe vs Temperature and Snow

TABLE III
Analysis of Temperature-Related Pipe Breaks
Recorded During the Most Severe Period on Record,
From December 1976 through February 1977

| Pipe Material | Years of Installation | Total Length, Ft | # Breaks Recorded | Breaks per Mile |
|---|---|---|---|---|
| Unlined DeLavaud Cast Iron | 1926-1937 | 350,000 | 274 | 4.13 |
| Asbestos-Cement | 1951-1966 | 7,600 | 4 | 2.78 |
| Unlined Sand Cast Iron | 1862-1929 | 1,210,000 | 94 | 0.41 |
| Cement Lined DeLavaud Cast Iron | 1934-1971 | 2,220,000 | 139 | 0.33 |
| Cement Lined Ductile Iron | 1964-1977 | 694,000 | 1 | 0.01 |
| Other | | 1,048,400 | 43 | 0.22 |
| All Pipes | | 5,530,000 | 555 | 0.53 |

Breaks are identified for each hazard scenario and inserted into the network model at nodal positions using probabilistic methods based on the break rate per length provided by the damage function and the total length of pipe supplied by the node. Historical data shows that the maximum number of damage breaks that should be assigned within a given region of the network model is limited to 10% of the number of buried pipe components in that region (Wang) for earthquake and 2% for cold temperatures (Cassaro).

The program permits the user two options for determining loss of individual components:

1. The user selects a damage condition for a component and obtains the hazard level and return period that will produce that damage.

2. The user selects a hazard scenario based on magnitude and return period and the program provides the expected damage to the system component.

For analysis of system loss, only the latter procedure is provided for in the program.

System Performance

Once system vulnerability has been determined and the loss estimates are obtained, system performance is analyzed using KYPIPE (Wood). Results of this analysis evaluates the system performance under the chosen hazard scenario by locating where results of damage cause water supply to be less than demand levels, or where pressure levels are below required pressure.

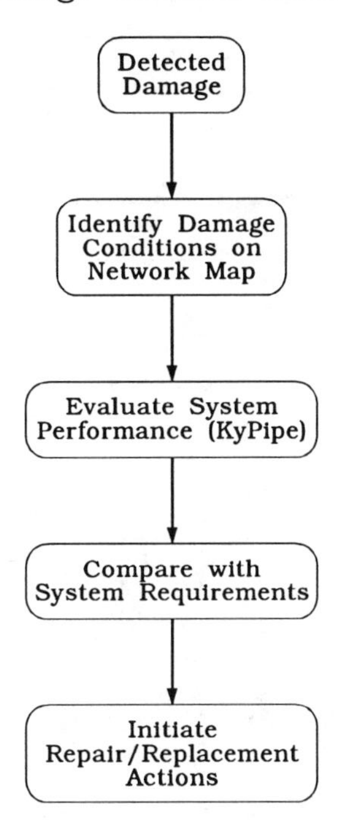


Figure 8 - Damage Control Model

Risk Analysis

For a defined set of hazard levels, the direct loss estimates due to repair and replacement costs for damaged components are represented as a probabilistic distribution as illustrated in Figure 9. The distribution of direct loss for a hazard level follows from the analysis of the losses estimated for the historical set of scenarios defined with the same hazard and intensity.

For a hazard scenario, the expected loss estimates due to fixed facilities are determined by,

$$E_f = DS_f * RV_f, \quad (1)$$

where $E_f$ is the estimated loss, $DS_f$ is the damage state determined for the facility, and $RV_f$ is the replacement value of the facility in dollars.

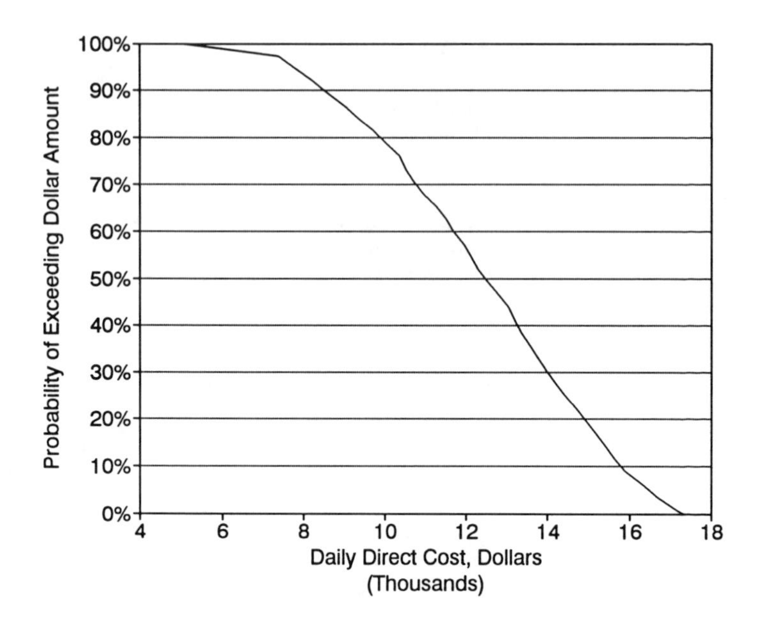


Figure 9 - Direct Cost of Pipe Breaks Due to a Temperature Hazard of 15° F and No Snow Cover (Shown in Figure 7)

The expected loss estimates due to pipe damage, $E_p$, are determined as a function of material and diameter. The summation of expected losses for all damage to the system represents the expected system loss under the hazard scenario,

$$E_s = \sum E_f + \sum E_p, \qquad 2$$

where $E_f$ is the estimated loss due to facility damage, and $E_s$ is the system loss in dollars.

Disaster Response

The program may be used within a water facility damage control center. Two applications are available:

1. As reports of structural damage or pipe leakage are gathered, the estimated damage is applied to the program, Figure 8. The expected system performance is computed and decisions made regarding repair and recovery.

2. Expected damage to the system may be estimated for a hazard event or in anticipation of a predicted hazard event.

## Summary

The program provides rapid risk assessment for water delivery systems subject to natural hazards. The risk management model is developed applying appropriate hazard conditions and specific water network topography to the system modeled. The risk assessment is used for planning and design management of the water delivery system. All data for defining natural hazards and expected damage are predetermined and placed in a knowledge base for direct application by water delivery users.

## Acknowledgement

This project is supported under grant No. BCS-8922660 by the National Science Foundation.

## References

An, D. (1991), "Interactive Effect of Structure-Pipe System Subjected to Ground Motion", M S Thesis, Department of Civil Engineering, University of Louisville, Louisville, Kentucky.

Applied Technology Council (1985), "Earthquake Damage Evaluation Data for California," ATC-13, Redwood City, California.

Army Corps of Engineers (1980), "Methodology for Area-Wide Planning Studies," EM 1110-2-502, U.S. Army Engineers Waterways Experiment Station, Vicksburg, Missouri.

Ballantyne, D.B., Berg, E., Kennedy, J., Reneau, R. and Wa, D. (1990), "Earthquake Loss Estimation Modeling of the Seattle Water System," Kennedy/Jenks/Chilton, Federal Way, Washington.

Baron, M. (1992), "Failure Mechanisms for Building Structures Under Lateral Forces," M Eng Thesis (in preparation), Department of Civil Engineering, University of Louisville, Kentucky.

Cassaro, M.J. and Huang, Y. (1992), "Effects of Severe Freezing Temperatures on Buried Pipes," Technical Report, Center for Hazards Research and Policy Development, University of Louisville, Kentucky.

Esmaeelzadeh, H.R. (1990), "A Preliminary Liquefaction Susceptibility Map for the Louisville Area," M Eng Thesis, Department of Civil Engineering, University of Louisville, Louisville, Kentucky.

Hansen, S.P., Gumerman, R.C., Culp, R.L. (1979), "Estimating Water Treatment Costs," Vol 2-3, EPA-6002-79-162 b, c, d, Cincinnati, Ohio.

Johnston, A.C. and Nava, S.J., (1985), "Recurrence Rates

and Probability Estimates for the New Madrid Seismic Zone," Journal of Geophysical Research, Vol 90, pp 6732-6753.
Khater, M.M , Grigorium M.D., and O'Rourke T.D. (1988), "Seismic serviceability of water supply systems", Proceedings of the 9WCEE, Tokyo-Kyoto, Vol VII, pp 123-128.
Liao, S.S.C., Veneziano, D. and Whitman R.V., "Regression Models for Evaluating Liquefaction Probability," Journal of Geotechnical Engineering Division, ASCE, Vol 114, p389-411.
Nuttli, O.W. and Herrmann, R.S. (1984), "Ground Motion of Mississippi Valley earthquakes," Journal of Technical Topics in Civil Engineering, ASCE, Vol 110.
Okumura, T. and Shinozuka, M. (1991), "Serviceability Analysis of Memphis Water Delivery System," Lifeline Earthquake Engineering, ASCE, TCLEE No. 4.
Schneider, E., Kamojjala, S., Burke, D. and Cassaro, M.J. (1992), "Expected Damage to Water Facilities for Natural Hazard Events," Report, Center for Hazards Research and Policy Development, University of Louisville, Kentucky.
Seed, H.B. and Idriss, I.M. (1971), "Simplified Procedure for Evaluating Soil Liquefaction Potential," Journal of Soil Mechanics & Foundations Division, ASCE, Vol 97, p1249-1273.
Seed, H.B., etal, (1984), "The Influence of SPT Procedures in Soil Liquefaction Resistance Evaluations," Report 190 UBC/EERC 84-15, Earthquake Engineering Research Center, University of California.
Taylor, C.E., (1991), Editor, "Seismic Loss Estimation for a Hypothetical Water System - A Demonstration Project," Technical Council on Lifeline Earthquake Engineering Monograph, ASCE.
Trautmann, C.H., O'Rourke, T.D., Grigoriu, M. and Khata, M.M. (1986), "Systems Model for Water Supply Following Earthquakes," Lifeline Risk Analysis Case Studies, ASCE, New York.
Wang, L.R.L., Wang, J.C.C., Ishibashi, I., Ballantyne, D.B. and Elliott, W.M. (1990), "Development of Inventory and Seismic Loss Estimation Model for Portland, Oregon Water and Sewer Systems," Report, Department of Civil Engineering, Old Dominion University, Norfolk, Virginia.
WAVES Program and Users Guide (1990), Earthquake Engineering Research Center, University of California, Berkeley.
Wood, D.J. (1991), "Computer Analysis of Flow in Pipe Networks Including Extended Period Simulations," Program and Users Manual, Department of Civil Engineering, University of Kentucky, Lexington.
Youd, L.T. and Bennett, M.J. (1981), "Liquefaction Site Studies Following 1979 Imperial Valley Earthquake," ASCE National Convention.

# Impact on Water Supply of a Seismically Damaged Water Delivery System

M. Shinozuka,[1] M. ASCE, H. Hwang,[2] M. ASCE, and M. Murata[3]

## ABSTRACT

This paper presents the damage evaluation of a water delivery system caused by the ground shaking in the event of an earthquake by using the LIFELINE-W computer program. The adequacy of post-earthquake water supply and the possible impacts on the residents and industries in an urban area are assessed. In this study, the Memphis water delivery system is used as an example. A scenario earthquake of 7.5 moment magnitude is assumed to occur at Marked Tree, Arkansas. If the supply nodes such as pumping stations and elevated tanks remain intact during the earthquake, it is predicted that the Memphis water network could deliver sufficient water to most areas in Memphis and Shelby County except a few census tracts. The post-earthquake water output in tract 202 is only about half of the intact condition. This is caused by severe ground shaking and poor network connectivity. Since this tract has a large population, shortage of water supply after the earthquake may have significant effects on the residents in this area. In addition,

---

[1] Sollenberger Professor, Department of Civil Engineering and Operations Research, Princeton University, Princeton, NJ 08544

[2] Professor, Center for Earthquake Research and Information, Memphis State University, Memphis, TN 38152

[3] System Analyst, PASCO Corporation, Tokyo, Japan

the water output is also reduced in tract 77 (President Island). The change in water supply may affect the industries on President Island.

## INTRODUCTION

People living in large cities are totally dependent on the lifeline systems that deliver water, gas, oil, and electricity. A lifeline system is usually very complex and covers a large area. If a lifeline system is damaged in the event of an earthquake, economic loss and impact on the society may be tremendous. Evaluating such an impact is quite challenging because of the complexity of a lifeline system. In recent years, an interactive and graphic computer program, LIFELINE-W, has been developed for evaluating the seismic performance of a water delivery system (Sato and Shinozuka 1991). One of the important features of this computer program is that it includes the ARC/INFO software package (Environmental Systems Research Institute 1988). ARC/INFO is a software based on the Geographic Information System (GIS) technology. Thus, the LIFELINE-W program has various functions such as digitizing a water delivery network, interactive editing of features and attributes, performing seismic risk analysis, and displaying results in graphic form. Furthermore, these functions are integrated by using the macro language and menu interface of the ARC/INFO program so that the computer code can be used easily and effectively. In this study, with the Memphis water delivery system as an example, the damage of the water delivery system in the event of an earthquake is evaluated by using the LIFELINE-W program. Then, the adequacy of post-earthquake water supply and the possible impacts on the residents and industries in the Memphis area are assessed.

## MEMPHIS WATER DELIVERY SYSTEM

The Memphis water delivery system is operated by the Memphis Light, Gas and Water Division (MLGW). The system supplies water in both the City of Memphis and Shelby County, Tennessee, except for a few incorporated municipalities inside the county. The water is taken from underground aquifer by deep wells and delivered to the public after treatment. Data pertinent to the entire Memphis water delivery system have been collected, digitized and implemented into a GIS data base to establish a realistic model of the water network (Figure 1).

The Memphis water delivery system consists of a large low-pressure system and several high-pressure systems on the outskirts of the city. The total length of the buried pipes is about 850 miles (1370 km) and the pipe diameter ranges from 6 inches (15 cm) to 48 inches (122 cm). The system is comprised of about 1300 links and 960 nodes. The node elevation varies from 210 ft (63.6 m) to 418 ft (126.6 m) and the average elevation is 296 ft (89.7 m). The unlined cast iron pipe was used for those installed before 1960, while the cement-lined cast iron and ductile iron pipes were used afterwards. As of 1990, there are eight pumping stations in the low-pressure system and one small pumping station in the high-pressure system. In addition to these pumping stations, six elevated tanks are used in the high-pressure systems and nine booster pumps connect the low-pressure and high-pressure systems.

According to statistics compiled by MLGW (1988), the water demand of the entire system varies from 119 to 179.2 million gallons per day (MGD) depending on the season. Under a normal operating condition, the water demand at each node can be determined. The nodes with high demand are mainly located inside Memphis city limits, while some are scattered throughout the county. The City of Memphis and Shelby County are divided into 133 census tracts. The water demands at the nodes within each census tract are integrated to establish the water demand for each census tract (Figure 2).

## FLOW ANALYSIS UNDER INTACT CONDITION

For the flow analysis of a water network, a computer code has been developed (Okumura and Shinozuka 1991). This code uses the analysis algorithm proposed by Takakuwa (1978). In addition, booster pumps and treatment of negative pressure are also included in the development of the code. For the flow analysis of the Memphis water delivery system, nine pumping stations and six elevated tanks are considered as supply nodes with fixed water head. These supply nodes are assumed not to be damaged by an earthquake. The vulnerability of the pumping stations is currently being evaluated. The effects of vulnerability

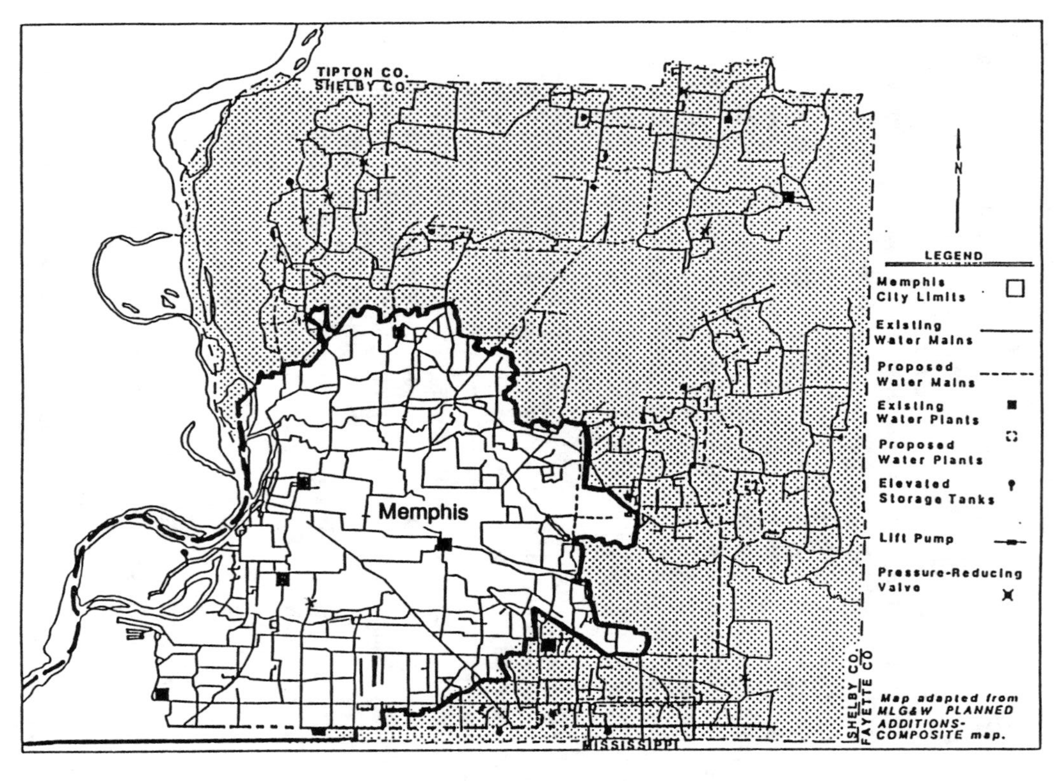


Figure 1. MLGW Water Delivery system

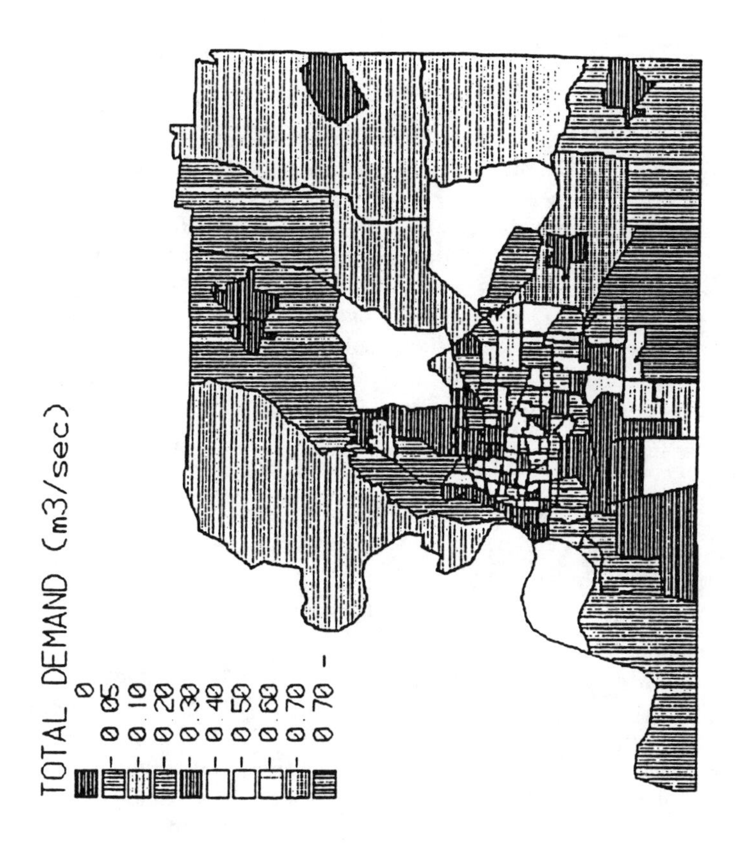


Figure 2. Water Demand in Census Tracts

of the pumping stations on the Memphis water delivery system will be investigated in the future. For the pipe attached to a booster pump, the water lift by the booster pump is added to the water head at the downstream nodes. The leakage at all the nodes is also taken into consideration. The water leakage at the node i is expressed as follows:

$$\lambda_i = C_i \, (E_i - G'_i)^k \tag{1}$$

where $\lambda_i$ is the water leakage in $m^3$/sec, $C_i$ is the leakage coefficient and is taken as 0.0002, $E_i$ is the water head in m, $G'_i$ is the node elevation in m, and k = 1.15 is used in this study.

A flow analysis of the Memphis water network under an intact condition is carried out to evaluate the flow parameters such as water head and output flow rate. The water heads at the nodes in the suburban area tend to be smaller than those inside Memphis, although the output flow rate can meet the demand at most of these nodes. The flow analysis results are laid over the census tract coverage to determine the total amount of output and the ratio of output to demand in each tract (Figure 3). The water supply is sufficient in all the tracts except in four tracts where the output is slightly less than the demand.

## SEISMIC GROUND SHAKING

Memphis is close to the southwestern segment of the New Madrid seismic zone (NMSZ). Thus, significant seismic hazards exist in the Memphis area. Hwang et al. (1990) evaluated the ground shaking in the Memphis area caused by a 7.5 moment magnitude earthquake assumed to occur at Marked Tree, Arkansas. The peak ground acceleration (PGA) values at 424 sites were determined by nonlinear site response analyses. In this study, the weighted moving average method (Burrough 1986) is used to interpolate these PGA values for the entire study area (Figure 4). Because of the location of the scenario earthquake, the ground shaking is more severe in the northwestern part of the county. However, there are some variations in the distribution of the PGA values reflecting the nonlinear soil effects on the ground motions.

The pipe failure data are usually established on the basis of the

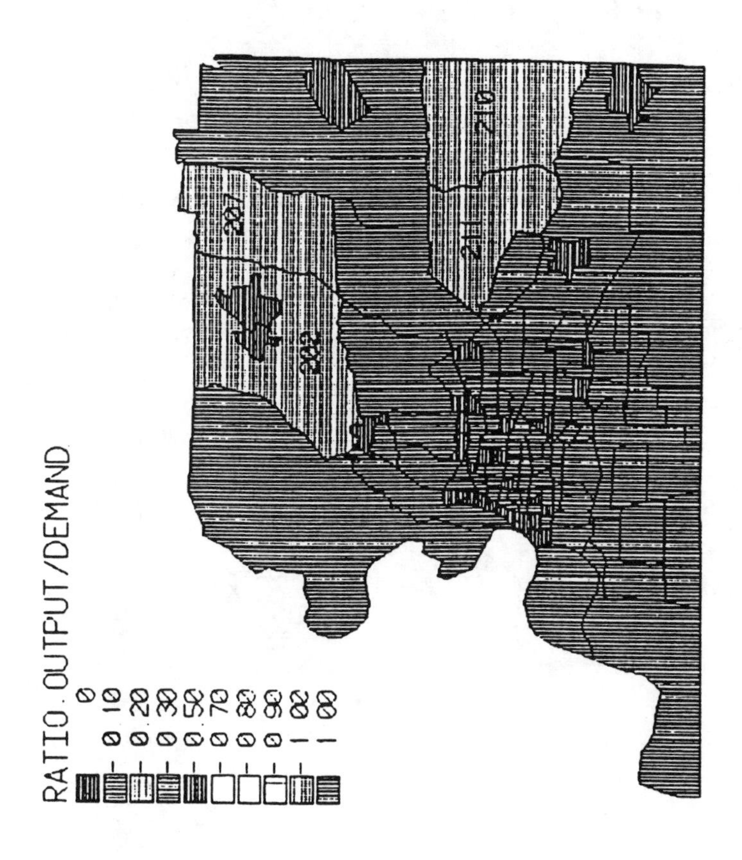


Figure 3. Water Supply in Census Tracts under Intact Condition

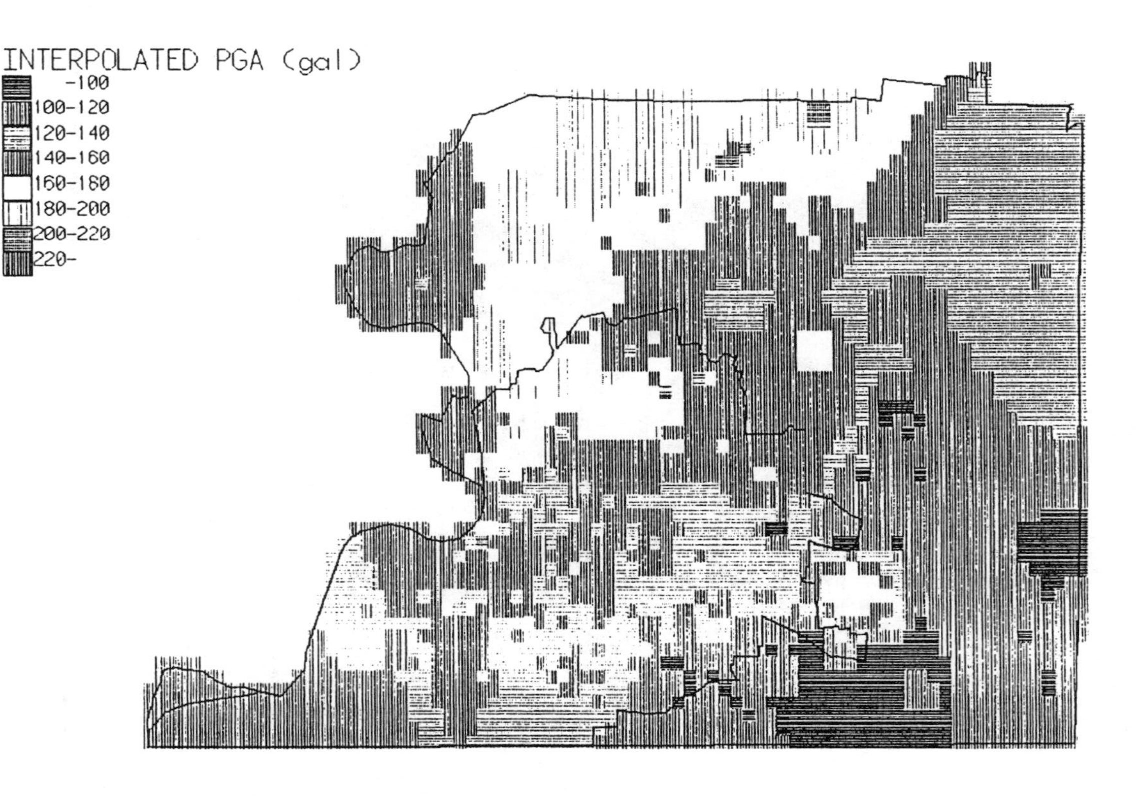


Figure 4. Peak Ground Accelerations in the Memphis Area

Modified Mercalli Intensity (MMI) scale. Thus, the PGA values are converted into corresponding MMI values by using the correlation relationship proposed by Gutenberg and Richter (1942). The MMI value in the study area (Figure 5) varies from VII-1/2 to IX with MMI of VIII-1/2 in the northwestern portion of the county and MMI of VIII in the southeastern portion of the county.

## DAMAGE STATE SIMULATION

The damage sustained by the water network caused by the ground shaking is simulated by using the Monte Carlo method. Table 1 shows the pipe failure rate corresponding to PGA or MMI intensity (Okumura and Shinozuka 1991).

**Table 1 Occurrence Rate of Pipe Failure**

| PGA (gal) | MMI scale | Pipe Failure Rate (break/km) | | |
|---|---|---|---|---|
| | | $D < 25$ | $25 \leq D < 50$ | $50 \leq D < 100$ |
| 30.5 | VI | 0.00398 | 0.00199 | 0.00080 |
| 65.3 | VII | 0.02512 | 0.01256 | 0.00502 |
| 139.6 | VIII | 0.15848 | 0.07924 | 0.03170 |
| 298.5 | IX | 1.00000 | 0.50000 | 0.20000 |

In Table 1, D is the pipe diameter in cm. By assuming that the occurrence of a pipe failure follows the Poisson process, the failure probability of the i-th pipe $P_i$ can be determined as

$$P_i = 1 - \exp(- r_i * L_i) \tag{2}$$

where $L_i$ is the pipe length in km and $r_i$ is the i-th pipe failure rate per unit length. The MMI intensity of the i-th pipe under the scenario earthquake is automatically assigned by laying the MMI intensity coverage over the water network. The failure rate

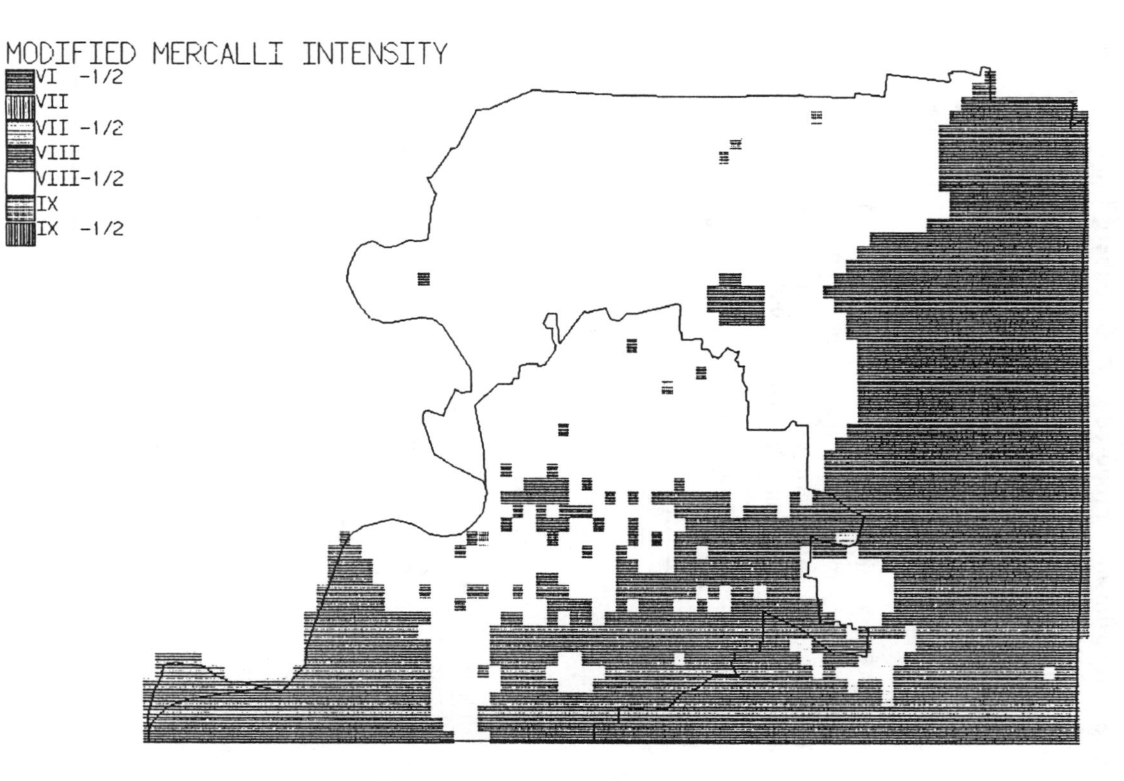


**Figure 5. MMI Intensities in the Memphis Area**

of the i-th pipe can then be determined according to Table 1. A random number uniformly distributed between 0 and 1 is generated for each pipe. If the random number is smaller than $P_i$, the pipe is considered to have been broken and thus it is removed from the network. Flow analysis of the damaged network is then performed to determine the water head and output flow rate at each node. On the basis of 25 simulation results, the average water head and the average output flow rate are determined for each node (Figure 6). Furthermore, the simulation results are laid over the census tract coverage to estimate the ratio of the output under damaged condition to the demand in each census tract (Figure 7). For most areas in Memphis and Shelby County, the predicted water output maintains at the level more than 80% of the demand; however, a moderate decline in water output occurs in the northern part of the county. The results reflect the stronger ground shaking as well as the poor connectivity of the network in this part of the county. This is especially true in tract 202, in which the output is only about half of the intact condition (53%). Considering the relatively large population in this tract, post-earthquake water use may be severely restricted because of the damage of the water network. Furthermore, tract 77 (President Island) has about three quarters of the normal water output (76%). Since President Island is an industrial area, post-earthquake water shortage may affect operation of the facilities in this area.

## CONCLUSIONS

This paper presents the damage evaluation of a water delivery system caused by ground shaking in the event of an earthquake by using the LIFELINE-W computer program. The adequacy of post-earthquake water supply and the effect of water shortage on the society are also assessed. This study demonstrates that the interactive and graphic computer program, LIFELINE-W, which includes the ARC/INFO package, is a very powerful tool for performing seismic risk analysis of a lifeline network. Spatial information such as the topography, census tracts, and intensity of ground motion can be easily integrated. In addition, the display of results in graphic form makes it easier for the user to examine analytical results and to make decisions effectively.

For evaluating the Memphis water delivery system, a scenario

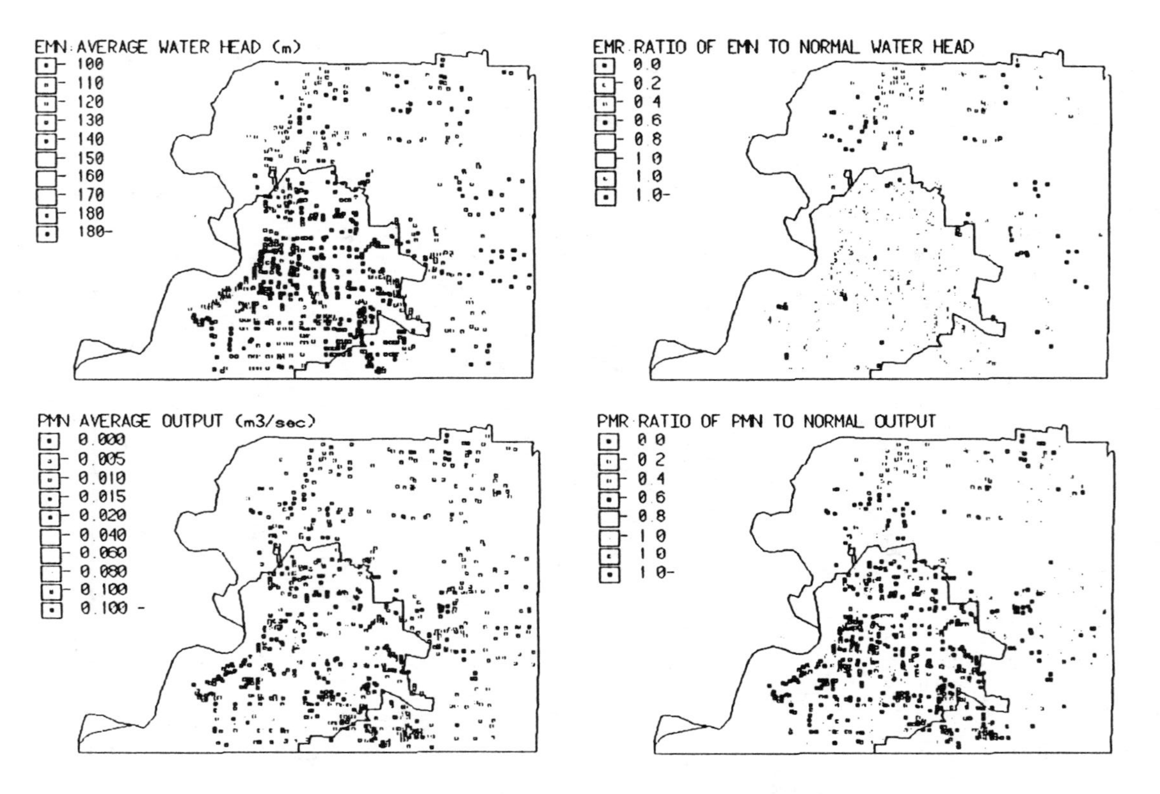


**Figure 6. Flow Parameters at Nodes under Damaged Condition**

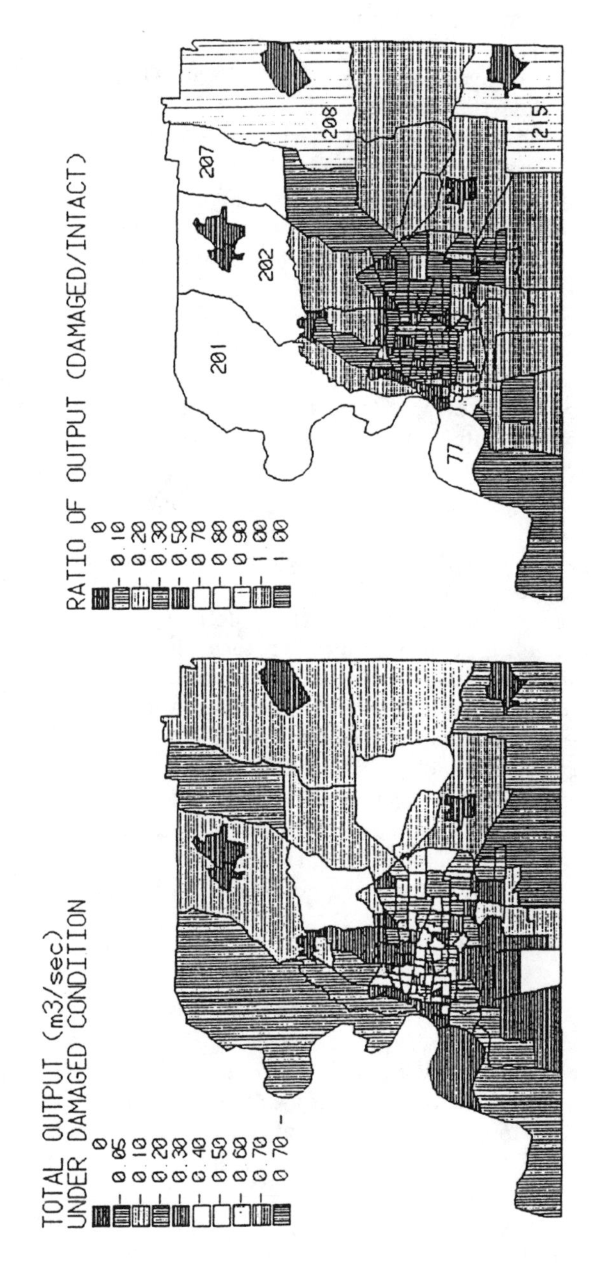


**Figure 7. Water Supply in Census Tracts under Damaged Condition**

earthquake of moment magnitude 7.5 is assumed to occur at Marked Tree, Arkansas. The damage state of the water delivery system is evaluated by using the Monte Carlo simulation. Under the assumption that the supply nodes such as pumping stations and elevated tanks remain intact during the earthquake and the pipe failure rate as shown in Table 1, it is predicted that the Memphis water network could deliver sufficient water to most areas in Memphis and Shelby County with the exception of a few census tracts. The post-earthquake water output in tract 202, which is in the northern part of the county, is only about half of the intact condition. This is caused by severe ground shaking and poor network connectivity in this area. Since this tract has a large population, shortage of water supply after the earthquake may have significant effects on the residents in this area. In addition, the water output is also reduced in tract 77 (President Island). The change in water supply may affect the industries on President Island.

In this study, the supply nodes such as pumping stations are assumed not to be damaged by ground motions. In fact, the pumping stations of the Memphis water delivery system may be quite vulnerable to ground shaking, because these facilities were not designed to resist earthquakes. In addition, broken pipes are assumed to be immediately closed by shut-down devices; therefore, the leakage from broken pipes are not considered. These assumptions need to be refined in future studies. Furthermore, the ground shaking is the only seismic hazard considered in the analysis. Other seismic hazards such as soil liquefaction may also have great potential to damage the network. Such a study will be undertaken in the future.

## ACKNOWLEDGMENTS

This work is supported by the PASCO Corporation, Tokyo, Japan, and by the National Center for Earthquake Engineering Research under contract numbers NCEER 903005 and 903009 (NSF grant number ECE-86-07591). Any opinions, findings, and conclusions expressed in the paper are those of the authors and do not necessarily reflect the views of NCEER or NSF of the United States.

The authors appreciate the generous assistance provided by the

Memphis Light, Gas and Water Division (MLGW), City of Memphis. In particular, the assistance given by Mr. Tommy Whitlow of MLGW is greatly appreciated.

## REFERENCES

Burrough, P.A. (1986), "Principles of Geographic Information Systems for Land Resources Assessment," Oxford University Press, New York, NY.

Environmental Systems Research Institute (1988), "ARC/INFO Users Manual," Redlands, CA.

Gutenberg, B. and Richter, C.F. (1942), "Earthquake Magnitude, Intensity, Energy, and Acceleration," Bulletin of the Seismological Society of America, 32(3), 163-191.

Hwang, H., Lee, C.S., and Ng, K.W. (1990), "Soil Effects on Earthquake Ground Motions in the Memphis Area," Technical Report NCEER-90-0029, National Center for Earthquake Engineering Research, State University of New York at Buffalo, Buffalo, NY.

Memphis Light, Gas and Water Division (1988), "Master Plan for Future Development of Water System," Memphis, TN.

Okumura, T. and Shinozuka, M. (1991), "Serviceability Analysis of Memphis Water Delivery Systems," Proceedings of the Third U.S. Conference on Lifeline Earthquake Engineering, Los Angeles, CA, 530-541.

Sato, R. and Shinozuka, M. (1991) "GIS-Based Interactive and Graphic Computer System to Evaluate Seismic Risks on Water Delivery Networks," Proceedings of the Third U.S. Conference on Lifeline Earthquake Engineering, Los Angeles, CA, 651-660.

Takakuwa, T. (1978), "Analysis and Design of Water Distribution Network (Haisui Kan-mou no Kaiseki to Sekkei)," Morikita Shuppan, (in Japanese).

**Seismic Hazards in the Eastern U.S. and the Impact on Transportation Lifelines.**

Klaus H. Jacob [1]

Abstract

In the eastern US *seismicity is low, hazard is moderate, and risk is quite high,* especially in metropolitan regions with vulnerable transportation systems typically not designed to resist earthquake loads. Seismic load ratios for earthquakes with 2,500-year vs. 500-year average recurrence periods measure from 2 to 4; hence, they are about 2-times higher in the eastern than in the western US. To account for these higher load ratios we suggest that for the design and retrofit of *ordinary* bridges, a recurrence period of about 1,000 years be chosen. For *essential* and *critical* bridges recurrence periods of about 2,500 and 5,000 years, respectively, are proposed (subject to further review). Ideally, design strategies should be fully risk-based (i.e. based on societal loss/cost/benefit relations), rather than mostly hazard-based as is essentially the case now. Site or soil factors in current seismic bridge design codes (e.g. AASHTO) do not serve well to account for the higher ground motion amplification on soft soils, nor reduction on hard rock, typical for the eastern US and for periods ≥.3sec. The site factors urgently need revisions for which models exist as examplified by the newly drafted New York City seismic code provisions. Quantitative methods need to be codified for screening soil profiles for their liquefaction potential. In the eastern US, soil liquefaction and slope instabilities have been observed to much larger distances than in the western US. In short: the seismic risk to bridges and transportation lifelines in the eastern US is higher than the low level of seismicity alone would suggest. This higher risk needs to be addressed for the design of new, and retrofit of existing systems. Risk- rather than purely hazard-based design/retrofit strategies could be cost-effectively applied to gauge the seismic safety requirements of essential and critical transportation systems which society expects to function even after significant earthquakes.

Introduction

The seismic hazard for the eastern United States is defined as the earthquake-related hazard of the region from east of the Rocky Mountains to the Atlantic coast. Most portions of the eastern US are seismically active (Figure 1). Features of earthquake hazards common to this region fundamentally differ from the seismic hazards in the western US. We elaborate on these differences.

With respect to transportation lifelines, we refer primarily to bridge and highway lifelines for which AASHTO (1983) has provided *guide specifications* for

[1] Senior Research Scientist, Lamont-Doherty Geological Observatory of Columbia University, Route 9W, Palisades NY 10964; Fax: (914) 359 5215; E-mail: jacob@lamont.ldgo.columbia.edu

seismic design. These and the *standard specifications* (AASHTO, 1989) are largely based on recommendations made originally by the Applied Technology Council for buildings (ATC-3, 1978), for design of new bridges (ATC-6, 1982), and for retrofit of existing bridges (ATC-6-2, 1983). These documents were based on data and experience largely from California, a region of higher level of seismicity and stronger ground motion attenuation than observed in the eastern US. Therefore they need critical review for their applicability in the eastern US, especially in the light of recent research results that have attempted to better quantify the seismic hazards in the eastern US.

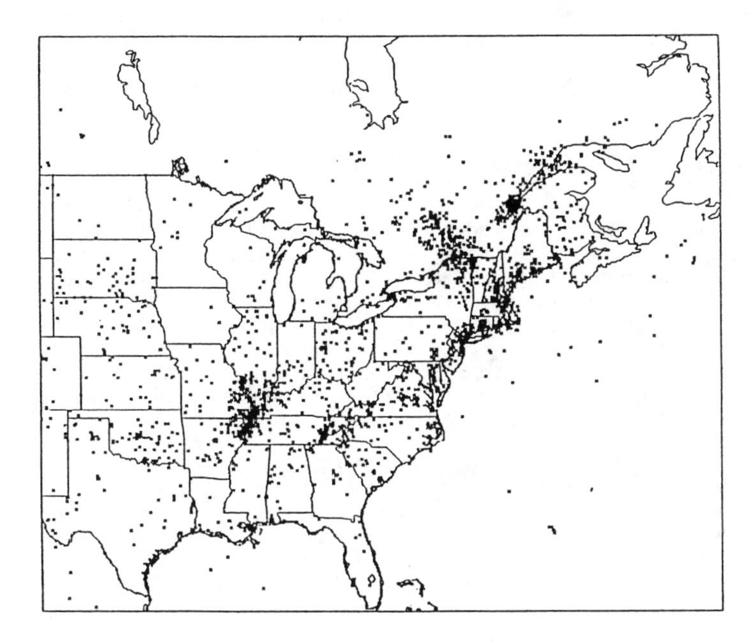

**Figure 1:** Seismicity (M≥3) for the eastern US and adjacent Canada.

We restrict our discussion of seismic hazard implication mostly to highway bridges. But the conclusions apply equally to other transportation and infrastructure systems, i.e. railsystems for which AREA has responsibility to set code guidelines, or airport systems that may fall under FAA and local regulations. Although we dont discuss the seismic safety of rail, airport, canal, tunnel or harbor systems, they do deserve more systematic attention in the future than they have attracted in the past.

The AASHTO seismic design provisions for highway bridges are presently under revision (Roberts and Gates, 1991). This revision is partly necessitated by introducing a new LRFD (Load and Resistance Fator)-based design concept. This effort is advanced under the auspices of the National Cooperative Highway Research Program's (NCHRP) Project 12-33. The revisions are not yet completed. Therefore it is timely to summarize some of the seismic hazard and risk concerns from an "Eastern" vantage point in the expectation that some of them can be further addressed in future revisions of AASHTO's code documents.

Current code revisions (Roberts and Gates, 1991) already address the following issues:

- the need for a forth soil category of deep soft soils, S4;
- a more explicit definition of structure importance (critical, essential, other);
- revision of response modification factors to account for structure importance;
- the necessity of considering longer-recurrence-period (2500-year) earthquakes for critical and essential bridges.

These revisions are important because, as we will see, they allow for design strategies that are more commensurate with eastern US earthquake hazard and risk environments. Some of the basic east-west differences regarding seismic risks are:

(1) *Nature of Seismic Hazard.* The hazard differs in the eastern US from that in the West because of:

- 10 to 100 times lower rates of seismicity (i.e. substantially longer average recurrence periods for the same magnitudes),
- lack of surface faulting (i.e. potential loci of future quakes are not well known),
- less rapid attenuation of ground motions with distance (i.e: larger damage areas),
- higher high-frequency content of seismic ground motions to large distances,
- higher contrast of shaking on soft soils vs. hard rock (high site amplification),
- liquefaction of non-cohesive soils appears to reach out to larger distances,
- the eastern ratio of ground motions between 2,500- and 500-year recurrence events can measure up to twice the ratio for the West. Given this trend, eastern bridges are likely to experience more severe damage from a 2,500-year event than western bridges, if both are designed for 500-year events. We will return to this important issue of larger eastern seismic load ratios.

(2) *Uncertainty in Quantifying the Hazard.* There is much greater uncertainty in quantitative hazard assessments for the East than in the West because:

- the historic seismicity record (300 years) is too short compared to the recurrence periods of major damaging events (thousands of years in the East). This leads to great uncertainties in estimates of the seismicity levels, of the spatio-temporal stationarity of the seismicity, and in the assignment of regional maximum magnitudes.
- only few sets of eastern strong-motion data exist. They marginally constrain the attenuation of shaking with distance, and dependence on magnitude.

These combined uncertainties can lead to high, more costly seismic design levels especially for critical bridges because for them the acceptable risk for damage is low. To avoid extended downtimes after earthquakes requires, however, that sufficiently high design (and retrofit) levels (for either strength or ductility) will be applied that are unlikely to be exceeded by future ground motions [alternate solutions are to install seismic isolation/damping devices to reduce the forces and motions allowed to enter into the superstructure (Buckle, 1991)].

(3) *Age and Fragility of the Built Inventory.* There exists a large stock of often aged eastern transportation systems (especially bridges) not originally designed to resist earthquakes. This bridge inventory is believed to be more vulnerable to earthquakes than most western systems that tend to be younger and are more often designed to resist earthquake forces.

(4) *Lack of Seismic Performance Record.* As a corollary to the large uncertainty in the seismic hazard, there exists a similar uncertainty about the seismic performance characteristics and fragility of eastern transportation systems (Mander and Chen, 1992) because during the systems' lifetimes few have sofar actually experienced significant seismic shaking.

(5) *Risk Concentration in Large Vulnerable Metropolitan Corridors.* The eastern US contains metropolitan corridors with the highest population densities in the US, with associated concentrations of traffic systems and high traffic volumes, especially along the Washington - New York - Boston corridor on the Atlantic coast. Significant portions of the corridor are seismically active. Therefore a single moderate earthquake (magnitude ≈6) can cause, e.g. in the New York City area, damage that could measure in the tens of billions of dollars, which together with indirect losses can approach a total loss for the national economy measuring up to a good fraction of a 100 billion dollars. How large this fraction actually could become strongly depends on the seismic performance of critical and essential transportation systems. The failure of the Bay Bridge between San Francisco/Oakland during the Loma Prieta earthquake provided an illustration from the west coast for the extraordinary socio-economic impact from the downtime of a single critical bridge in a metropolitan area. While population concentrations exist of course in the western US, i.e. Los Angeles, San Francisco, or Seattle, they are not linked to the at least perceived higher seismic fragility of the eastern built environment.

These few introductory remarks may serve as a warning that the relatively low level of seismicity of eastern regions translates into moderate hazard, and more importantly, into relatively high potential seismic risk. It is in this low-seismicity/ high-risk environment that we need to develop and implement risk-sensitive strategies for seismic design of new bridges and transportation systems, and perhaps even more important, for seismic retrofit of the existing vulnerable structures.

## Quantifying Seismic Hazard as a Function of Recurrence and Exposure Periods.

*Seismicity - Entire Eastern US.* Most portions of the eastern US are seismically active (Figure 1), although the levels of seismicity are only about one tenth to one hundreth of the seismicity rates in the most active areas of California or Alaska. Let us first estimate the probability of occurrence of earthquakes anywhere in the eastern US, East of the Rocky Mountains. Using a seismicity catalog for the entire eastern US (e.g. Seeber and Armbruster, 1991) we can determine

$$\log n = A - bM \qquad [1]$$

In this equation, $n$ is the cumulative average number of earthquakes (normalized to one year and for the entire eastern US) with magnitudes ≥M. The average recurrence period, for events with magnitudes ≥M, is then T=1/n. Let us assume the events are all part of a random Poisson process (they occur independently of each other). Then we can define the probability P(%) that during an exposure time Δt (which may be taken as the expected lifetime of a new structure, or as the expected remaining lifetime of an existing structure) an event with magnitude M, corresponding to an average recurrence period T(M), will occur. This probability is:

$$P(\%) = 100\,(1 - e^{-\Delta t / T(M)}) \qquad [2]$$

Using actual numbers derived from the quoted eastern US earthquake catalog, we compute the average recurrence periods and probability values given in Table 1.

Table 1: Probabilities of Earthquakes in the US East of the Rocky Mountains:

| Magnitude M | Rec. Period T(years) | Probability P(%) that event M will occur* in time Δt(years) Δt=10y | 20y | 30y | 50y | 100y | 1,000y |
|---|---|---|---|---|---|---|---|
| 5 | 3 | 98 | ≈100 | ≈100 | ≈100 | ≈100 | ≈100 |
| 6 | 22 | 37 | 61 | 75 | 90 | 99 | ≈100 |
| 7 | 180 | 5.5 | 10 | 16 | 24 | 43 | ≈100 |
| 8 | 1,470 | 0.7 | 1.4 | 2.0 | 3.3 | 6.6 | 49 |

* probabilities are for the occurrence of events in the *entire* US region east of the Rocky Mountains without distinction exactly where the events would occur.

The computed average recurrence periods T are large, especially for the larger magnitudes (M≥7), when compared to the short times lapsed since such events have actually occurred last during the historic record. For example, the M≈7 Charleston S.C. earthquake of 1886 occurred only 106 years ago, while the average T for an M≈7 event to occur anywhere in the eastern US based on the over-all seismicity is computed to be about 180 years; the M≥8 New Madrid earthquakes of 1811/12 occurred only about 180 years ago, while the average recurrence period for such an event to occur anywhere in the eastern US is estimated to be T≈1,470 years. We see that, on average, truly large events are expected to be quite rare in the eastern US, provided that extrapolations from past seismicity are a valid guide for future seismicity. This crucial assumption of seismic stationarity cannot be proven with great confidence because of the short historic record. Stationarity of seismicity is simply a convenient and commonly adopted working hypothesis, but should be applied with the greatest possible caution. It does not provide conservative margins of safety in regions of past low seismicity. Too often designers and planners ask for the nearest known fault or most significant historic earthquake, and if none are on record they feel they can minimize consideration of the earthquake risks. This is poor judgment.

*An Example: Seismicity of the NY City Metropolitan Region.* We next use the historic earthquake catalog (≈300 years), amended by local network data for the last 20 years, for a region centered on New York City. Moreover, we make the assumption of long-term spatial homogeneity in the level of the random seismicity around New York. With this assumption one can estimate equally likely events, described as pairs of magnitude and distance (Jacob, 1990), for given annual probabilities $p_a$ of their occurrence. We select two levels, $p_a$ =0.2% and $p_a$ =0.03% per year. They correspond to average recurrence periods of approximately T≈500 and T≈3000 years. We could have chosen any other values. The results for these two probability levels, or average recurrence periods, are summarized in Table 2.

Table 2 illustrates a basic relation that holds only probabilistically, that is, it holds *on average, while actual individual events may stray far from these averages for any given locality and selected recurrence period.* For a preselected annual probability of occurrence, or a preselected average recurrence period, we find that the larger the magnitude to which we expect to be exposed, the larger is (on average) the distance at which the earthquake can be expected to occur. Conversely, the longer the average recurrence period which we consider (or the lower the annual probability of occurrence that we select as an acceptable hazard level), the nearer one can expect the event to occur, given a preselected magnitude. Or, given a preselected distance, the expected magnitude increases with increasing recurrence period, or decreases with increasing annual probability of occurrence. It is important to restate again that these are probabilistic statements. The very fact that we assume that earthquakes of virtually any size can occur somewhere in the eastern US implies that there will always be some localities (bridges or cities) that are closer to larger earthquakes than those quoted in Table 2. Yet values of the kind shown in this table are the ones which are typically selected when developing design and retrofit ground motion input levels. It follows, that design events have a finite chance to be exceeded. This chance for exceedance becomes lesser, however, the longer the average recurrence period T that is adopted for design purposes.

Table 2: Probable Distance d(km) from NY City for Earthquakes with Magnitude M

| Magnitude M | Expected Distance d(km) from New York City | |
|---|---|---|
| | for $p_a$ =0.2% or ≈10% chance in 50 years or T≈500 years | for $p_a$ =0.03% or ≈10% chance in 300 years or T≈ 3000 years |
| 4 | 16km | 7km |
| 5 | 46 | 19 |
| 6 | 132 | 54 |
| 7 | 490 | 200 |
| 8 | 1400 | 570 |

*Seismic Load-Ratios: Eastern Seismic Loads do not as Readily Saturate with Long Recurrence Periods.* Given the diffuse and poorly defined seismicity of the eastern US, how does the seismicity translate into levels of expected ground motions and seismic loads for structures? And how do the seismic loads increase with the recurrence periods of seismic events? These are societally crucial questions because the public has a strong interest in whether especially the *critical* and *essential* structures will be able to survive some fairly large nearby earthquakes, i.e. those with long average recurrence periods.

To compute the ground motions associated with various recurrence periods we combine the seismicity information contained in Table 2 for equally probable magnitude-distance events, with the attenuation laws for how spectral ground motion parameters (e.g. 5% damped response spectral acceleration $S_a$ or velocity $S_v$) vary with epicenter distance and magnitude. For this purpose we apply attenuation laws for ground motions measured in the Eastern US to the seismicity information of Table 2 and obtain approximate Uniform Hazard Spectra, UHS (Figure 2), as described elsewhere (Jacob, 1990). A UHS is constructed in such a way that all its spectral ordinates share equal probabilility of exceedance. Because of this 'uniform hazard' property, UHS are often used as a basis for code design spectra. The level and shape of each UHS (Figure 2) depends on the chosen recurrence period T, in addition to depending on the seismicity and attenuation laws characteristic for the region. Examples for the two UHS shown in Figure 2 apply to ground motions on very hard crystalline rocks (defined in the NYC code as $S_0$), typical for New York City (Jacob, 1990).

Note in Figure 2 that the amplitude ratio of the two uniform hazard response spectra which represent a recurrence-period ratio of about six (3000y vs. 500y), varies as a function of the natural period of the structure. The spectral amplitude ratio measures about 2 for the acceleration-flat portion of the spectra (for periods ≤.3 sec); it is about 3 for the velocity-flat portion (from 0.3 to 2 sec); and approaches 4 for the displacement-flat portion (≥ 2 sec). These results are consistent with the notion that the long-period (displacement-flat) portions of the spectra grow rapidly (in fact linearly) with the increasing seismic moments of the larger earthquakes (located at shorter distances) for increasing recurrence periods T; while the accelerations grow only slowly with increasing magnitudes, despite the decreasing distance of the events as recurrence period T increases.

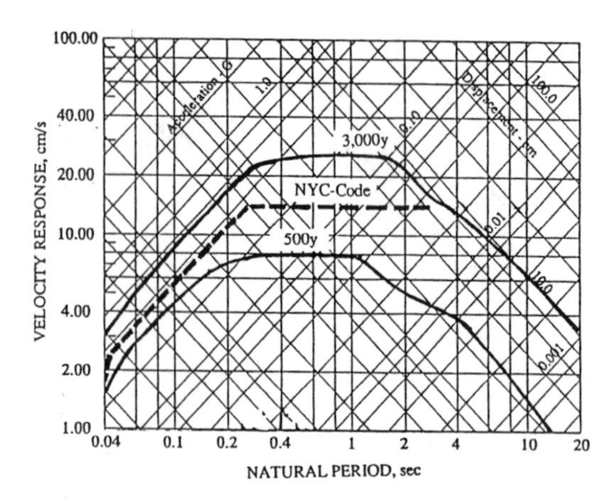


**Figure 2:** Uniform-hazard response-spectra for two recurrence periods (500 and 3,000 years), and the building code design spectrum for NY City. All spectra apply to hard-rock sites ($S_0$).

*Comparison with Seismic Load Ratios from NEHRP Maps.* Recent response-spectral acceleration maps were prepared by Algermissen et al. (1991) for the 1991-edition of the NEHRP seismic provisions (FEMA-BSSC, 1992). They show smaller and less frequency-dependent spectral load ratios than those just described, partly because they correspond to a smaller recurrence period ratio of about 5 (10% exceedence for 250 vs. 50 years) instead of 6. The NEHRP maps were prepared for two natural periods, 0.3 and 1 sec. The maps are referenced to $S_2$ (deep stiff soil) conditions, in contrast to the $S_0$ hard-rock conditions for the New York uniform hazards spectra (Fig. 2). The spectral ratios for the 10% exceedance in 250 vs. 50-year maps of Algermissen et al. (1991) average around 2.2±0.3 for high-seismicity, high-acceleration California sites, for both 0.3 and 1.0 sec periods; for lower-seismicity, lower acceleration sites from Utah to the Atlantic coast, the ratios measure 2.5±0.3 and 3.0±0.3 for 0.3 sec and 1.0 sec natural periods, respectively.

From these maps we confirm our earlier observation that the spectral load ratios are higher in the eastern US (low-seismicity regions) than they are for the western

US-California (high-seismicity regions). However, the Algermissen et al. (1991) ratios (2.5 at 0.3 sec, and 3.0 at 1 sec) for the eastern US do not as strongly increase with structural period as those derived for the eastern US from the UHS values presented in Fig 2. Given some of the underlying differences in methods and attenuation laws used by these authors, the load ratios (3±1) for the pair of exceedance probabilities of 10% in 250 vs. 50 years, agree surprisingly well, especially for ground motion periods of about 1 sec and less. The increased, frequency-dependent load ratio typical for the eastern low-seismicity regions has important consequences for the seismic design levels of large critical eastern bridges with large structural periods (≥1 or 2 sec), especially if some of their foundations happen to be imbedded in deep soft soils overlying hard bedrock. An example for such a structure on soft soils is the Tappan Zee bridge carrying the NY Thruway (I-287) across the Hudson River, just north of New York City.

Geologic Site Amplification and Soil Coefficients for Eastern vs. Western Seismic Environments and Codes.

Sofar we have restricted the discussion to how ground motions and seismic load ratios vary with average recurrence period, while assuming standard reference site conditions. Even with these limitations, and apart from the fundamental differences in level of seismicity, we already found diverging trends between eastern and western US hazard parameters. Site amplification can further increase these differences. Recent efforts to modify site factors used in codes on a national scale (Whitman, 1992) and for the eastern US (Jacob, 1990, 1991) are ongoing. Some basic unresolved issues related to soil/site amplification are:

- (1) Poor (non-quantitative) definitions of soil or site categories ($S_1$, $S_2$, $S_3$, $S_4$).
- (2) Coarseness of soil depth intervals. Thinner soil layers must be accounted for.
- (3) The range and magnitude of soil amplification factors are not large enough to cover soil/rock impedance contrast conditions for the eastern US (Jacob, 1991). Also non-linear soil response in the eastern US cannot be as readily invoked for adopting low soil amplification factors (as suggested by experts for the western US) because the input ground motions are, on average, likely to be lower in the East than in the West.

Item (1), *Soil Profile Definitions:* The descriptions of the soil categories ($S_1$, $S_2$, $S_3$, $S_4$) currently used in AASHTO (or for that matter in seismic building codes like NEHRP, UBC, BOCA, SEAOC) are qualitative instead of quantitative. For example, they do not use shearwave velocity or standard penetration test (SPT) blow counts to characterize the soils geotechnically. Instead imprecise descriptive soil terms are used to identify soil profiles. Quantitative geotechnical descriptors are urgently needed. Models for such quantitative descriptions are presented in Jacob (1990) and discussed in Whitman (1992).

Item (2), *Coarse Depth Intervals:* The depth intervals permitted by the current code definitions are too coarse and therefore do not serve well in seismic environments where the high-frequency (short-period) content of ground motions is already high and will be further amplified by thin surface layers of soils over hard rock. Except for soft clays, thin layers are largely ignored in existing (westcoast-based) codes. As a rule of thumb, for seismic code purposes and for structures with fundamental natural periods of $T_0 \leq T^*$, any soil layer of thickness $H^* \geq T_0 v_s^* / 4$ should be considered which gives rise to an amplification of at least $A^* = (I_r/I_s^*)^{1/2} \approx (v_r/v_s^*)^{1/2} \geq 1.5$ at the fundamental soil-layer resonance period

$T^*=4H^*/v_s^*$. The meaning of the different variables used here is (Jacob, 1991): I is the elastic impedance (shear wave velocity v times density); subscript r and s refer to rock and soils, respectively; $v_s^*$ is the depth-weighted average shear wave velocity of the soil layer with thickness H*. To give an example: Present AASHTO (or national building) code definitions of soil categories do not differentiate a site with 50 feet of sands or silts with an average shear wave velocity of 500 ft/s overlying hard bedrock with a shear wave velocity of ≥8,000 ft/s from a site with bare bedrock, although the soil site would have an (elastic) site amplification of at least 4 [i.e. $(8000 / 500)^{1/2}$] at periods $T \leq 4H^*/v_s^* = 4 \times 50 / 500 = 0.4$ sec compared to the bedrock site. Such high amplifications (of 4) at periods of engineering consequence (0.4 sec) are conditions often encountered in the eastern US, yet would fall under $S_1$=1 site conditions in existing codes; therefore the soil categories presently used in the codes are quite useless and need to be changed urgently (Jacob, 1990; 1991).

Item (3): *Insufficient Range and Magnitude for Soil/Site Amplification Factors.* Site amplification factors for past AASHTO specifications were restricted to 3 soil-profile categories ($S_1$ through $S_3$) with numerical values for the S-factors limited to 1.0 to 1.5. Recent updates to AASHTO add a $S_4$=2.0 category for deep soft clays. Nevertheless, the very limited numerical range in amplification factors between 1 and 2 is inadequate especially for eastern US conditions for at least two reasons:

(i) *Linear vs. Non-Linear Soil Response:* lower average ground motion excitation levels typical for the East, e.g ≤0.2g peak acceleration on stiff ground, even when propagated into softer overlying soils, do not lead as readily to non-linear soil response as do the higher peak design accelerations of ≥0.4g typical for the more active portions of California. Therefore, in the East, the nearly full elastic amplification due to the elastic impedance contrast between soil and rock does, on average, become effective. In contrast, the soil factors adopted for the western US based on local experience, although not nearly big enough to fit actually observed site amplifications, are justified to be lower than in the low-excitation environment of the eastern US. During a given exposure time, non-linear soil response has a higher chance to occur in high-seismicity regions (of the West) than in low seismicity regions (of the East).

(ii) *Impedance Contrasts:* rarely can one find shear velocities of near-surface rocks in California as high as one does find them in the glacially eroded cratonic or Paleozoic portions of the eastern US where hard-rock surface velocities can reach 10,000 ft/s. Contrast this with soft organic clays with shear velocities of 300 ft/s directly overlying these rocks in some coastal estuaries, and one obtains impedance contrasts exceeding values of 30, and amplification factors of at least 5 or 6. It is for this reason that in deviation to national codes NY City, and recently NY State, have introduced into their newly drafted seismic codes 5 soil/rock categories with the numerical values $S_0$=2/3 for very hard rock; and for inrcreasingly softer and deeper soil sites $S_1$=1.0 ; $S_2$=1.2; $S_3$=1.5; and $S_4$=2.5 (for details see Jacob, 1990, 1991). This option allows for a much larger range of site amplification factors of $(S_4/S_0)$=3.75, compared to $(S_4/S_1)$=2 or $(S_3/S_1)$=1.5 in older codes based on western US experience. We strongly recommend that AASHTO follow the general lead of these recent code drafts to make its bridge code more suited for eastern US applications. Except for the New York Codes, most building codes (SEAOC, UBC, BOCA, NEHRP, Massachussetts, Standard/Southern BC) must face the very same problems.

Liquefaction Potential.

Soil liquefaction occurs because of reduction or loss of shear strength of non-cohesive, poorly consolidated, water-saturated soils (mostly sands and silts) when pore pressure builts up during and some time after earthquake shaking. Liquefaction can last for many minutes, locally even for hours, after an earthquake and may result in major permanent ground deformations. Friction piling may loose its lateral constraint and sink, deform, buckle or break, especially where differential lateral or vertical movements between soil layers are involved. Bridge foundations, retaining walls and abutments, earthen ramps, embankments, dams, and flat-lying or sloped made land (fill), when insufficiently compacted, are prone to liquefaction effects from earthquakes. Case studies of liquefaction occurrence, large ground deformations, and their effects on engineering structures have been compiled in an instructive series of NCEER workshop proceedings (O'Rourke and Hamada, 1989, 1991, 1992; Hamada and O'Rourke, 1992, Jacob and Turkstra, 1989).

Observed liquefaction features in eastern North America are limited in number but some of them are among the most widespread that have been documented on Earth (Obermeier, 1984; Obermeier et al., 1989). Most conspicuous among them are the M≈7 1886 earthquake in Charleston SC, and three New Madrid earthquakes in 1811/12 of magnitudes M≥8 followed by many strong aftershocks. This event sequence liquefied many soils in the Mississippi Valley in an area measuring 200 km by 50 km, straddling the five central US states of Arkansas, Tennessee, Kentucky, Missouri and Illinois. Liquefaction from lesser earthquakes (M≈ 5 to 6) has been recently documented in some detail in the northeastern US and in adjacent Canada (Tuttle and Seeber, 1989; Tuttle et al. 1990).

Youd and Perkins (1987) developed a Liquefaction Severity Index (LSI) ranging from 1 (low) to 100 (high) in order to qualitatively describe and classify the effects of soil liquefaction. Comparing the attenuation of the LSI with epicentral distance, and as a function of magnitude, Youd et al. (1989) find that, for a given magnitude and distance, higher LSI values are observed in the eastern US than in the West. This implies that for a given magnitude the same LSI value is observed in the East at a larger distance than in the West. It is equivalent to saying that a *smaller* earthquake can cause liquefaction in the East that would require a *larger* event in the West. For example, during the M=5.9 1988 Saguenay earthquake in Quebec, Canada, embankment failures ocurred at distances of 140 and 170 km (!!) from the epicenter (EERI, 1988). At the distant site, a 50-ft high railway embankment failed, probably liquefaction-induced. The track was left suspended over a length of 300 ft, and the rail line was closed for one week for repairs. This type of long-distance effect for such a moderate earthquake is virtually unthinkable in the western US.

Despite the potential risks from liquefaction and other seismically induced ground deformations, most bridge and building seismic codes are deficient in addressing this hazard in a quantitative way. The issue is largely left to individual engineering judgement. Only the Massachussetts and the drafted NY City and State building codes use quantitative *screening* procedures that force the engineer to check for possible liquefaction hazards. The Commentary (Part 2), but not the Provisions (Part 1) of the 1991 NEHRP model code published by FEMA/BSSC (1992) addresses liquefaction, but the comments are almost entirely based on west-cost experience. The NEHRP document needs to be reviewed to account for eastern conditions more adequately. AASHTO should consider accelerated drafting of quantitative code guidelines for liquefaction and ground deformation with consequences for bridge foundations, embankments, abutments and retaining walls.

Conclusions: Societal Implications and Options for Eastern Design-Strategies.

*First,* when one discusses how the occurrence of earthquakes may effect design strategies for lifelines and other man-made structures, one must differentiate between three quantities: seismicity, seismic hazard, and seismic risk. Seismicity describes the recurrence rates of earthquakes with different magnitudes. Hazard quantifies the recurrence rates of certain ground motions (and of secondary effects like liquefaction or land slides). Risk, however, describes the potential for damage, destruction, or losses (both in terms of lifes and $-values, including indirect losses that society will incur from the lack of service after an earthquake). Risk is the spatially and temporally integrated product of: hazard, value of assets, and fragility of assets. In the eastern US these three quantities (seismicity, hazard, and risk) can in general be qualitatively characterized as: *low* seismicity, *moderate* hazard, and *high* risk. This relation may be illustrated by a somewhat hypothetical example: The M=7 Loma Prieta earthquake of 1990 near San Francisco caused a $ 7-billion loss and the failure of one critical and several essential bridge/highway structures. Let us assume (without claiming any accuracy) that such an event occurs, on average, about once every hundred years in the SF Bay Area. Near New York City, a magnitude M≈7 earthquake may occur, on average (if at all), only once every 1,000 years, or even less frequently. But when it does, it would cause losses almost surely exceeding $ 70-billion (in 1990-dollars). Also it would probably put out of service for some time about half a dozen critical bridges and probably some tunnels, scores of essential, and hundreds if not thouands of ordinary bridges, not to speak of human losses. Annualizing the two losses, $7billion/100y vs. $70billion/1000y yields a similar estimated loss rate of $70million per year, while seismicity rates are at least a factor 10 (and perhaps almost a factor 100) lower in NY than in the SF Bay Area. It is for this reason that we state the seismic risk potential is *relatively* high, i.e. relative to the rate of seismicity, and even more so, relative to the public *awareness* of either seismicity, hazard, or risk.

*Second,* for the designer or owner of transportation lifelines, it is important to realize that if one designs for 500-year recurrence time events in the SF Bay Area, one has designed almost for the largest credible earthquakes (M≈8), since in that time period the largest events may have occurred, on average, about two times. Designing for the 2,500-year recurrence time events does in this high-seismicity-region not increase much the *level* of ground motions that one would experience, because essentially only the number of occurrences, but not the level of motions would much increase. Not so in the East! A 500-year event (in NY City a M≈51/2 event) is far from a maximum credible event of, say, M≈7 (the magnitude of the Charleston S.C, or the Grand Banks, Canada, earthquakes). Therefore, designing for a 500-year event leaves in New York City (or the East in general) a lower margin of safety against a 2,500-year event, than when designing for 500-years recurrence in San Francisco. Because in the East the seismic loads increase more rapidly with rising recurrence periods, thus creating higher load-ratios, one can question whether it is too risky to follow in the East the AASHTO guideline of using seismic zone factors consistent with 500-year recurrence periods. While these guidelines were adopted nationally, they were based mostly on California experience, aimed at buildings rather than bridges. We note that the New York City code design spectra for *ordinary buildings* (which may not always be so ordinary by national standards) are already more consistent with a 1,000- rather than a 500-year recurrence period (Figure 2, and Jacob [1990]). CALTRANS does not follow the AASHTO approach either; it retains a deterministic approach based on *maximum credible spectra.* For the more distant future we suggest that instead of basing the design input motions purely on a uniform hazard (whether 500- or 1000-years), and then modifying them for the importance category of the bridge, a perhaps more rigorous approach would be to quantify the total risk/cost/benefit relation and optimize design or retrofit strategies accordingly by aiming at the highest total-risk-based benefit/cost ratio. This more sophisticated approach may be justified only for essential and critical structures. It requires substantial socio-economic regional data input. Also a codified standard procedure for this approach still needs to be developed and adopted by expert and public consensus. We also refer to a novel approach taken for the TCA Bridge Study (unpublished report, 1990) for three planned major toll roads in the LA area, Orange County, CA. In that study different design strategies are compared and their cost implications evaluated. Until new concepts fully emerge, we suggest that in the East *ordinary* bridges be designed for 1,000-year-recurrence events, *essential* bridges for ≥2,500-year events, and *critical* bridges for ≥5,000-year events.

*Third,* for the design of transportation lifelines attention needs to be directed to the relatively high potential for soil liquefaction, large ground deformations, and their effects on bridge foundations and earthen embankments if poorly consolidated soils are present. Eastern soil conditions, site-amplification and high-frequency spectral content together appear to enhence the liquefaction potential.

*Fourth,* for large bridges with large foot prints and spaced multiple supports, the differential ground motions, and their coherence (or lack thereof), need to be carefully considered. At longer periods it is often not the dynamic accelerations (forces), but the differential displacements between two points of support that may be critical for the survival of a bridge. The motions of a stiff superstructure may exceed the limited seat width of a joint or bearing, thus causing the super-structure to drop unless it has additional restrainers that can prevent this to occur. 500-year events in the East tend to have a lower displacement-to-acceleration ratio than western 500-year events. In the East, this is good news with respect to limited seat width. But the bad news is that the rate with which this displacement-to-acceleration ratio increases with increasing recurrence period in a low- vs. a high-seismicity region, is higher in the East than in the West. Therefore, to introduce reasonable margins of safety against events with large recurrence periods, seat width and other system features sensitive to displacements (rather than forces or stresses/strains) need to be designed to either accomodate the rapidly increasing displacements, or to restrain the relative motions from occuring in the first place (for retrofit purposes CALTRANS, for instance, has used cable restrainers or other limiting devices).

*Fifth,* because ground motions in the East are enriched in short-period (high-frequency) energy (compared to the West), it may be advantageous to consider designing new bridge structures, or retrofitting existing ones, with base-isolation/damping devices. They have the tendency to shift the natural period of the structure to longer periods and to absorb energy by increased damping. This reduces the dynamic forces especially in small stiff structures because they move the response of the structure into the longer-period portion of the acceleration spectrum with expected lower amplitudes. But the engineer needs to be aware that increased seat width may be required to accomodate the larger displacements at the longer periods. This solution of using isolation-dampers may not be as beneficial for large structures and/or structures on soft soils, for which already high-amplitude long-period response is expected. To shift the structural period into a high-amplitude portion of the ground-motion spectrum would defeat the intended purpose of reducing the input motions by placing the fundamental structural period in a low-amplitude portion of the site-pecific design spectra. Careful site-specific site-response studies are called for to avoid such possible pitfalls.

Acknowledgments: This work was supported by grants from the National Center of Earthquake Engineering Research (NCCEER) with joint funding by the National Science Foundation and the New York State Science and Technology Foundation. I appreciate comments by Ian Buckle and thank Ron Mayes for bringing the TCA bridge study to my attention.

## References

AASHTO (1983), Guide Specifications for Seismic Design of Highway Bridges, American Assoc. of State Highway & Transportation Officials. Washington D.C.

AASHTO (1989), Standard Specifications for Highway Bridges, American Assoc. of State Highway & Transportation Officials. 14th Edition, Washington D.C.

Algermissen et al. [8 authors] (1991), Probabilistic ground-motion hazard maps of response spectral ordinates for the United States. *Proceedings, 4th Intern. Conf. Seismic Zonation,* held at Stanford CA; 8/25-29/1991; *EERI,* Oakland, CA., Vol II, pp. 687-694.

ATC (1978), Tentative Provisions for the Development of Seismic Regulations for Buildings. Applied Technology Council Report ATC 3-06 to the National Science Foundation and National Bureau of Standards. Redwood City, CA.

ATC (1982): Seismic Design Guidelines for Highway Bridges. Applied Technology Council Report ATC-6, Redwood City, CA.

ATC (1983): Seismic Retrofitting Guidelines for Highway Bridges. Applied Technology Council Report ATC 6-2, Redwood City, CA.

Buckle, I. (1991), Principles of seismic isolation. ASCE Structures Congress '91 Compact Papers, pp. 651-655. ASCE, New York NY.

EERI (1989), The Saguenay, Quebec, Canada, earthquake of 25 Nov. 1988, EERI Special Earthquake Report, EERI Newsletter Vol. 23 No.5, May 1989, pp.8.

FEMA/BSSC (1992), NEHRP recommended provisions for the development of seismic regulations for new buildings, 1991 Edition, Part 1 (Provisions) and Part 2 (Commentary, Chapter 7, pp.147-170.), FEMA/BSSC, Washington DC.

Hamada,M. and T.D.O'Rourke [edit.] (1992), Case studies of liquefaction and lifeline performance during past earthquakes; Volume 1: Japanese case studies; Techn. Report NCEER-92-0001,

Jacob K.H. and C.J. Turkstra [editors] (1989), Earthquake hazards and the design of constructed facilities in the eastern United States, Annals of the New York Academy of Sciences, Vol. 558, pp. 457, New York.

Jacob, K. (1990), Seismic hazards and the effects of soils on ground motions for the greater New York City metropolitan region. *In:* Geotechn. aspects of seismic design in the N.Y.C. metrop. area.; Risk Assessment, Code Requirements and Design Techniques. Metrop. Section, ASCE, NY NY, Nov. 13/14, 1990. pp. 24.

Jacob, K. (1991); Seismic Zonation and Site Response: Are Building-Code Soil-Factors Adequate to Account for Variability for Site Conditions Accross the US? *Proceedings, 4th Intern. Conf. Seismic Zonation,* held at Stanford CA; 8/25-29/1991; *EERI,* Oakland, CA., Vol II, pp. 695-702.

Mander J.B. and S.S. Chen (1992), Seismic performance of highway bridges in the eastern US, NCEER Bulletin Vol. 6, No. 2, April 1992, pp. 1-6; Buffalo NY.

Obermeier S.F., R.E. Weems, R.B. Jacobson, and G.S. Gohn (1989), Liquefaction evidence for repeated Holocene earthquakes in the coastal region of South Carolina. *In:* Jacob K.H. and C.J. Turkstra [editors] (1989), Earthquake hazards and the design of constructed facilities in the eastern United States, Annals of the New York Academy of Sciences, Vol. 558, pp. 183-195.

Obermeier S.F. (1984), Liquefaction potential for the Mississippi Valley. U.S. Geol. Survey Open File Report 84-770. pp 391-446.

O'Rourke T.D. and M.Hamada [editors] (1989), Proceedings from the second US-Japan workshop on liquefaction, large ground deformation and their effects on lifelines. Techn. Report NCEER-89-0032, Buffalo NY.

O'Rourke T.D. and M.Hamada [editors] (1991), Proceedings from the third Japan-US workshop on earthquake resistant design of lifeline facilities and countermeasures for soil liquefaction. Techn. Report NCEER-91-0001, Buffalo NY.

O'Rourke T.D. and M.Hamada [editors] (1992), Case studies of liquefaction and lifeline performance during past earthquakes; Volume 2: U.S. case studies; Techn. Report NCEER-92-0002,

Roberts J.E. and J.H. Gates (1991), Seismic design provisions of new LRFD code. ASCE Structures Congress '91 Compact Papers, pp. 486-489. ASCE, New York NY.

Seeber, A. and J.G. Armbruster, The NCEER-91 Earthquake Catalog: Improved Intensity-Based Magnitudes and Recurrence Relations for U.S. Earthquakes East of New Madrid. Technical Report NCEER-91-0021. Natl. Center for Earthqu. Engin. Res.; Buffalo NY.

TCA Bridge Study (1990), Development of Project-Specific Seismic Design Criteria. Report prepared for: Corridor Design Management Group, by Cumputech Engineering Services, Imbsen & Associates, and Woodward-Clyde Consultants.

Tuttle, M.P. and L. Seeber (1989), Earthquake-induced liquefaction in the northeastern US: historical effects and geological constraints. In: Jacob K.H. and C.J. Turkstra [editors] (1989), Earthquake hazards and the design of constructed facilities in the eastern United States, Annals of the New York Academy of Sciences, Vol. 558, pp. 196-207.

Tuttle, M., K. Tim-Law, L. Seeber and K. Jacob (1990), Liquefaction and ground failure induced by the 1988 Saguenay, Quebec, earthquake. Canad. Geotech. J. Vol. 27, pp. 580-589.

Whitman R.V. [editor] (1992), Proceedings from the site effects workshop, Oct. 24-25, 1991. Techn. Report NCEER-92-0006, Buffalo NY.

Youd T.L. and D.M.Perkins (1987), Mapping of liquefaction severity index. ASCE Journal of Geotechnical Engineering, Vol. 113, No. 11; pp 1374-1392.

Youd T.L., D.M.Perkins and W.G.Turner (1989), Liquefaction severity index attenuation for the eastern United States. *In:* O'Rourke T.D. and M.Hamada [editors], Proceedings from the second US-Japan workshop on liquefaction, large ground deformation and their effects on lifelines. Technical Report NCEER-89-0032, Buffalo NY, pp.438-452.

# EVALUATION OF SEISMIC VULNERABILITY OF HIGHWAY BRIDGES IN THE EASTERN UNITED STATES

J. B. Mander[1], F.D. Panthaki[2] and M.T. Chaudhary[2]

**ABSTRACT**

For high risk seismic zones such as California, the state-of-the-art in earthquake resistant design for new highway bridges has been well advanced, especially since the damaging 1971 San Fernando earthquake. The catastrophic failures in the 1989 Loma Prieta earthquake brought a new awareness to the vulnerability of existing non-seismically designed bridge structures, particularly those in the eastern U.S. Instead of applying code-based formulations as an inverse to the design process to determine the vulnerability of existing bridges, this paper presents an energy-based evaluation methodology. Results are presented for a one-quarter scale model gravity load designed pier, which show that in spite of poor detailing, gravity load designed structures can possess a high degree of intrinsic lateral strength and ductility capacity.

## HISTORICAL BACKGROUND

The catastrophic failures that occurred in Californian bridge structures during the 1971 San Fernando earthquake led to a considerable amount of sponsored research in which fundamental problems relating to the seismic design of bridges were studied. Much of this work was sponsored by the Federal Highway Administration (FHA), the National Science Foundation (NSF) and other agencies. As a consequence of the ensuing investigations the Applied Technology Council (ATC), which had previously developed guidelines for the seismic design of buildings, performed

---

[1]Asst. Prof., [2]Grad. Student, Department of Civil Engineering, State University of New York at Buffalo, NY 14260, U.S.A.

the task of developing similar guidelines for bridge structures. This resulted in the subsequent publication of ATC-6 (1979). Later the American Association of State Highway and Transportation Officials (AASHTO, 1983) recommended that ATC-6 be adopted as a "Guide Specification". The use of ATC-6 remained optional in most states within the U.S. until 1991 when the provisions were adopted into the AASHTO specifications.

In addition to the above mentioned research efforts undertaken in the U.S., considerable work has been done in New Zealand. Based on governmental and university research work the N.Z. National Society for Earthquake Engineering published in 1980 a series of papers concerned with the seismic design of bridges (NZNSEE., 1980). That work plus subsequent efforts, now forms the basis of seismic design practice for both road and railway bridges in New Zealand (see for example; Mander et. al. (1984), and Priestley and Park (1987)).

By virtue of the fundamental nature of these past research efforts, a better understanding now exists for the seismic design and response of ductile bridge substructures (piers) as well as the span seating and bearing requirements of bridge superstructures. Although considerable research progress has been made since the 1971 San Fernando earthquake many questions remain unanswered. Past research efforts have principally concentrated on the seismic *design* of new bridges in regions of high seismic activity, such as California. Only a limited research effort has been directed to the evaluation and seismic rehabilitation of the numerous bridge systems that already exist.

The two joint U.S.-N.Z. workshops, sponsored by NSF, held to systematically identify research needs, resulted in combined researcher and practitioner opinions on research priorities. These have been documented in ATC-12 (1981) and ATC-12-1 (1985). Such workshops are generally attended by engineers well versed in the practice of seismic design for the most adverse (Californian) circumstances. Only more recently has the problem of existing non-seismically designed structures, such as in the eastern U.S., received more widespread attention. However, as discussed by Poland et al (1989), this attention has initially focused on buildings. As a result of the number of bridge failures and deaths in the October 1989 Loma Prieta earthquake (in Spectra 1990), there is now a widespread awareness amongst practitioners outside California of the problems associated with existing bridge structures not engineered for seismic resistance. The particular motivation for extending the present state-of-the-art is to enable the large stock of existing bridges to be reliably evaluated for seismic risk, and if necessary rehabilitated.

**CAPACITY DESIGN VERSUS EVALUATION FOR REHABILITATION**

Some of the principal differences resulting from seismic design of bridge piers are manifest in the reinforcing detailing, as shown in Fig. 1. It is the implicit objective of seismic detailing, resulting from the capacity design of the structural system, to supply a high degree of ductility capability through flexural hinging as well as averting undesirable failure mechanisms such as shear, and bond and anchorage failures. If however, the pier does not possess all of these desirable attributes, it does not necessarily imply that a brittle failure will result for the following two reasons:

(i) It is not possible to infer the lateral load capacity and ductility capability of an existing gravity load designed bridge as a direct inverse problem to the seismic (capacity) design of a new structure. This is because capacity design is unable to predict the behavior of undesirable failure mechanisms: it only seeks to avoid them.

(ii) Seismic detailing practices as required by codes, has been derived from a vast body of experimental and theoretical research. This has attempted to maximize the ductility capacity of structural members for applications in the most adverse seismic environment. Performance inferences for structures with limited ductility demand, such as bridges in the eastern U.S., may be unreliable when based on present body of data and/or code requirements.

In eastern U.S. bridges, abutment bearings and girder seats are often considered to be the most vulnerable elements in earthquakes. In high risk seismic zones, current practice would require retrofitting with isolation bearings and providing shear keys or restrainers to limit seat width demand. However, retrofitting the vulnerable superstructure may only move the problem to the substructure. The piers of many eastern U.S. bridges have not been designed for seismic resistance and exhibit a non-ductile post-yield behavior. Retrofit methods which principally add ductility to piers are currently under development in California (refer to Chai et.al. (1991)). There is a need to develop appropriate measures of improving the ductility capacity of bridge columns in low to medium seismic zones where the demands are more modest.

If either the deck seating and/or the piers are retrofitted, it may leave the foundation system vulnerable. This could include spread footings and pile foundations as

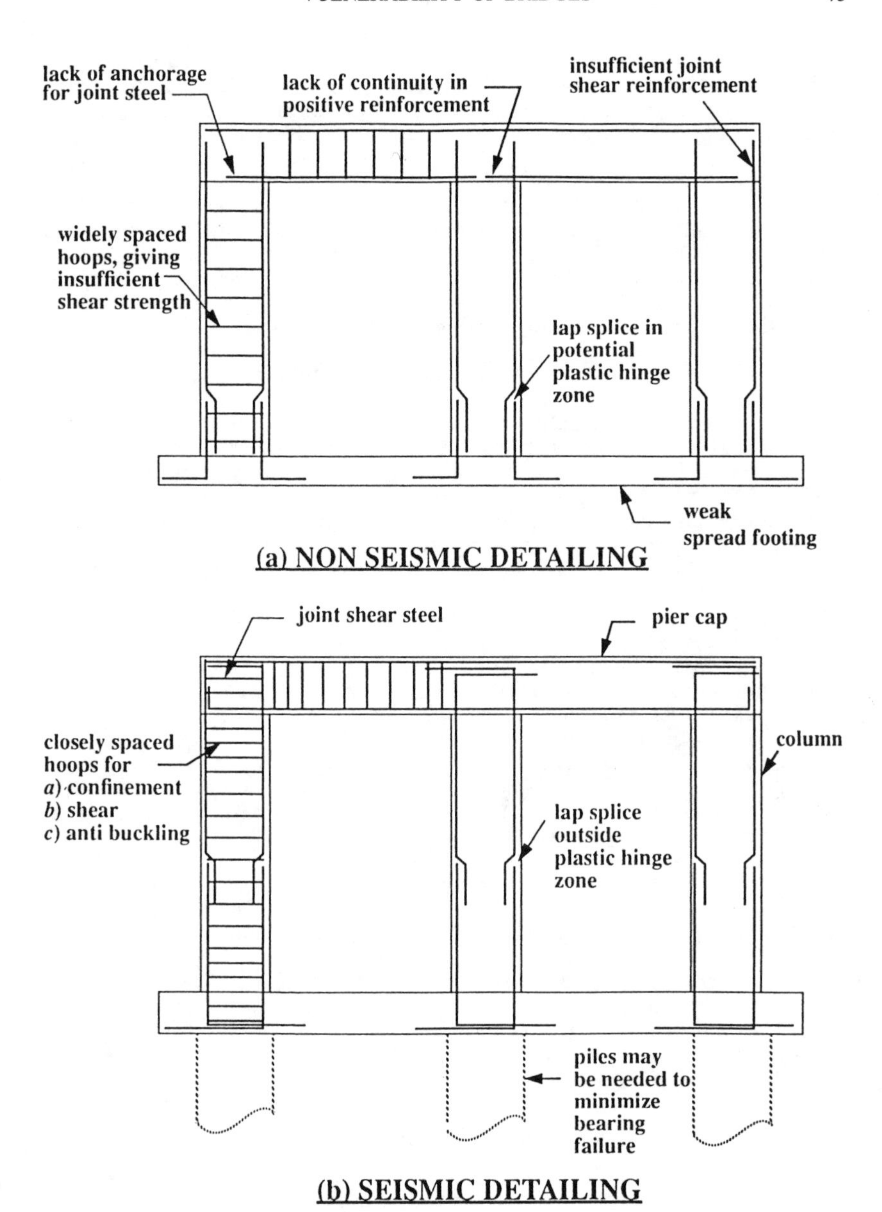


FIG. 1: Differences in the Reinforcing Detailing of Bridge Piers as a Result of the Design Philosophy.

well as the abutments. These components in a bridge may be more tolerant to damage than the piers or deck bearings. However, the damage potential of the foundation system is clearly important and needs to be definitively assessed.

The remainder of this paper proposes a seismic evaluation methodology which is based on balancing the expected energy input (demand) from a design earthquake motion, with the required energy dissipation capacity of the structural system. An example is then presented of a one-quarter scale model gravity load designed pier that exhibits poor detailing for shear resistance.

## VULNERABILITY EVALUATION OF EXISTING BRIDGE PIERS

The way in which most existing eastern U.S. gravity load designed bridges could fail in the event of a strong earthquake is not well understood. An analyst evaluating the seismic damage potential of a bridge structure needs to determine the hierarchy of failure mechanisms before viable retrofitting measures can be designed. Current evaluation techniques may not capture all of the possible modes of failure. For example, analytical techniques are currently not capable of capturing the level of loading required to collapse steel bridge bearings, or to cause columns that are inadequately reinforced for shear to fail. A clear need therefore exists to identify, in terms of local seismicity and structural type, the probability of failure. That is, both the return period of a catastrophic earthquake, as well as the return period of a damaging earthquake (that may render a bridge system unserviceable due to a partial failure) need to be determined. The role soil-structure interaction plays in bridge dynamics also needs to be included in the evaluation of existing structures. Theoretical models, in the form of transient non-linear dynamic analysis, and also seismic limit analysis approaches need further development.

In the present research, attention is focused on measures of hysteretic damage, rather than traditional notions of ductility. Structural damage at time t, during a seismic excursion, can be defined in terms of an energy-based damage index:

$$D(t) = \frac{E_h(t)}{E_c} \tag{1}$$

in which $D(t)$ = the accumulated damage up to time $t$, $E_h(t)$ = absorbed hysteretic energy to time $t$, and $E_c$ = total energy absorption capacity to failure. When $D(t) \geq 1$ then incipient failure (collapse) occurs. Alternatively, an identical result at the end of a seismic event can be obtained by

$$D_t = N_D/N_c \quad (2)$$

in which $D_t$ = total damage at the end of the event, $N_D$ = the number of full reversed cycles of demand determined from an analysis similar to that described below, and $N_C$ = the number of cycles (capacity) that an element can sustain to failure (at the same displacement amplitude to the demand cycles).

For a given bridge structure and appropriate ground motion the damage potential evaluation procedure may be summarized in the following three-step process:

(1) Evaluate the hysteretic DEMAND.

(2) Evaluate the hysteretic energy CAPACITY.

(3) Evaluate the Damage Index for the prescribed ground motion. If $D(t) > 1$ for the target ground motion the pier should be seismically retrofitted.

The manner in which hysteretic demand and capacity can be evaluated is explained in the following two sections. These remarks pertain particularly to bridges in the eastern U.S.

**EVALUATING HYSTERETIC ENERGY DEMAND**

In order to determine the ductility capacity required by the critical plastic hinge regions of bridge piers, the hysteretic energy demand must firstly be assessed. This may be computed for an entire structure using non-linear time history programs. The analysis should use the appropriate hysteretic models for member behavior.

For simple SDOF type structures such a sophisticated level of analysis is not warranted. Instead, hysteretic energy and cyclic loading spectra can be used. A study has been undertaken in which hysteretic energy spectra are derived for given ground motions using a non-linear dynamic time-history analysis. The method used in this study is a modified form of the absolute energy formulation proposed by Uang and Bertero (1990). This approach considers the accumulating components of energy resisting the total input energy from the earthquake, thus

$$E_i = E_k + E_s + E_\xi + E_h \quad (3)$$

in which $E_k$ = kinetic energy in the system, $E_s$ = strain energy, $E_\xi$ = absorbed viscous damping energy, and $E_h$ = hysteretic energy absorbed by the structure.

Fig. 2 shows the various spectra for an earthquake motion. Elasto-plastic single-degree-of-freedom (SDOF) systems were considered for different levels of force (strength) reduction $R$, defined as:

$$R = S_a/C_d \tag{4}$$

in which $S_a$ = normalized spectral acceleration required for an elastic response, and $C_d$ = non-dimensional base shear strength coefficient.

The most significant application of this analysis is to use the hysteretic energy and displacement response spectra to derive the equivalent number of fully reversed elasto-plastic cycles. For an elastic response ($R$=1) the number of cycles $N_d$ is defined as

$$N_d = \frac{\Sigma|X|}{4S_d} \tag{5}$$

where $S_d$ = spectral displacement, and $\Sigma|X|$ = the total displacement for the duration of the motion. Note that the minimum value of $N_d$ is equivalent to a monotonic response and is equal to 0.25.

The demand spectra results for the 1940 El Centro N-S earthquake record presented in Fig. 2 show that $N_d<4$ for all possible natural periods and strength reduction factors. Similar analyses have been carried out for other earthquake ground motions and hysteretic models not shown here. Results from these analyses show similar trends: that values of $N_d$ for short and medium periods are generally constant for a specific $R$ value. For example when $R$=6, $N_d$ was found to be approximately equal to 5, and 0.8 for the 1985 Mexico City and San Salvador 1986 earthquakes. These earthquakes respectively represent low-frequency/long-duration and high-frequency/short-duration events. This energy based analytical approach, together with realistic ground motions for the eastern U.S., can now be used to evaluate the energy based DEMAND on bridge substructures and components. This step is essential when attempting to plan and evaluate a realistic set of experiments.

One of the inherent advantages of the proposed energy-based methodology is that it implicitly accounts for the duration of ground shaking through the equivalent number of equi-amplitude cycles of motion. Ductility is also implicitly considered through the energy absorption demand.

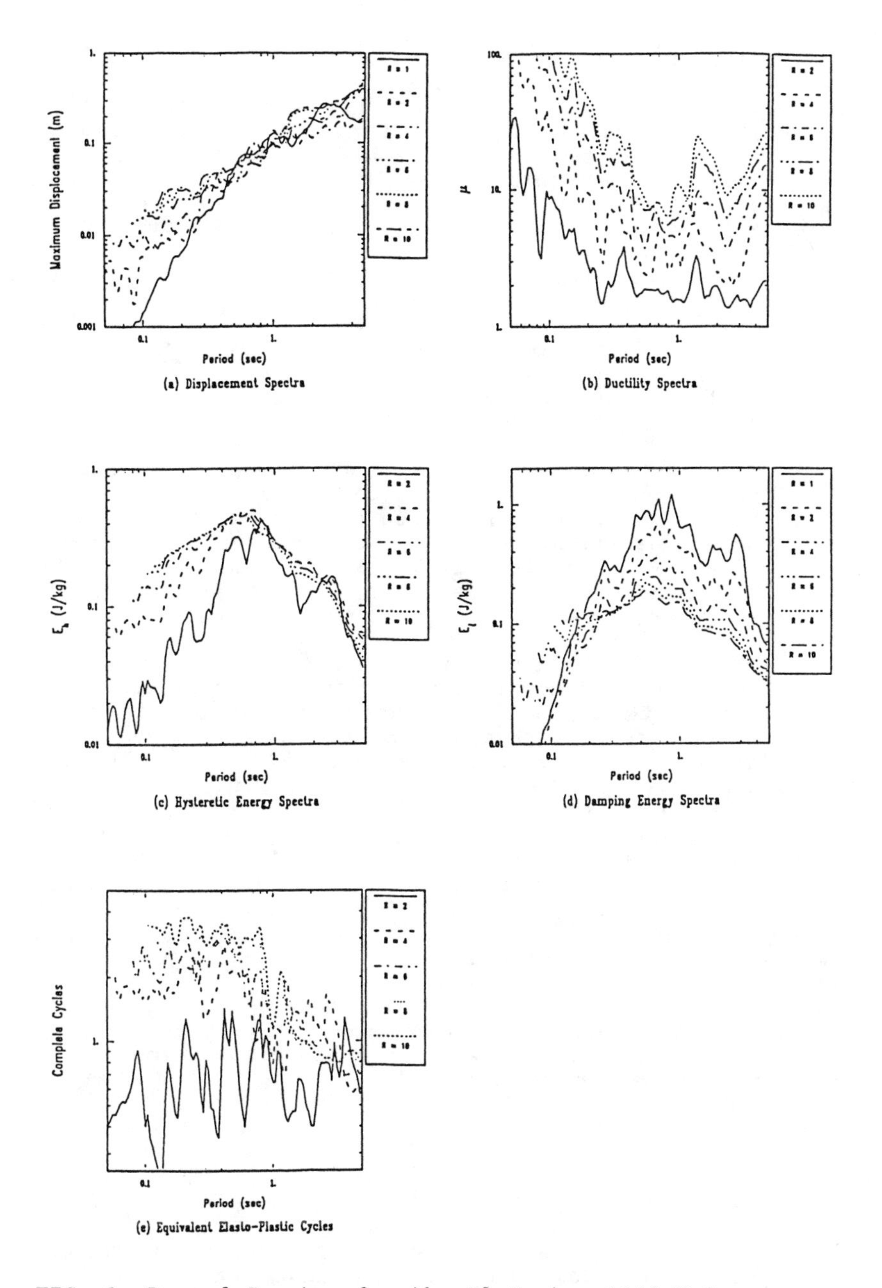


FIG. 2: Demand Spectra for the El Centro 1940 N-S Earthquake

**EVALUATING HYSTERETIC ENERGY CAPACITY**

It is important when assessing the hysteretic energy capacity of a system to have a clear definition of failure. For bridge piers failure can be defined as that state in which the pier is unable to sustain the gravity load of the superstructure, that is, the onset of collapse. Incipient pier collapse may result from the following failure types:

(a) The longitudinal bars fracture due to low cycle fatigue.

(b) The transverse hoops fracture, thus leaving the column unconfined. (Although this is not a collapse state, strength deterioration is very rapid following hoop fracture.)

(c) The lateral capacity is reduced to zero due to either shear strength deterioration, P-delta effects, or both. (In previous studies, investigators have sometimes defined this state as when the moment capacity reduces to say 80% of the nominal strength. Note that this may define a ductile serviceability limit state, but not a collapse state.)

The evaluation of hysteretic energy capacity should be carried out using analytical moment-curvature and/or force-displacement techniques. For such an analytical evaluation, reliable inelastic mechanistic micro-models are necessary to capture cyclic loading behavior. To date such models have been developed for flexurally dominated piers in which the shear span (M/VD) is greater than about 2.0. For flexural piers with low levels of axial load governed by type (a) failure, the strain range amplitude ($\epsilon_a$) of the longitudinal steel in alternating cycles can be determined from a simple monotonic moment-curvature analysis. An example problem is presented in Fig. 3 and the analysis procedure is described below.

Recently Panthaki (1991) tested high alloy prestressing threadbars and ordinary mild steel deformed reinforcing bars under low cycle fatigue conditions. From his results shown in Fig. 3(a) it is evident that a dependable (lower bound) fatigue relation can be described by the following equation, and can be used to determine $N_f$, the number of cycles to failure:

$$\epsilon_a = 0.08\ (2N_f)^{-0.5} \tag{6}$$

Fig. 3(b) shows the results of a moment curvature analysis for an example column with axial load intensities ($P/f'_c A_g$) of 0.0, 0.1 and 0.3. Note that this column has been

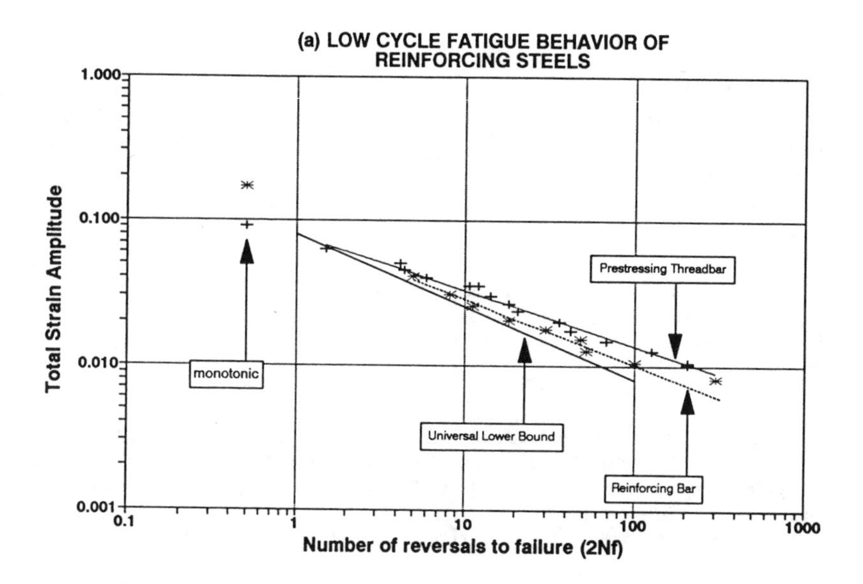


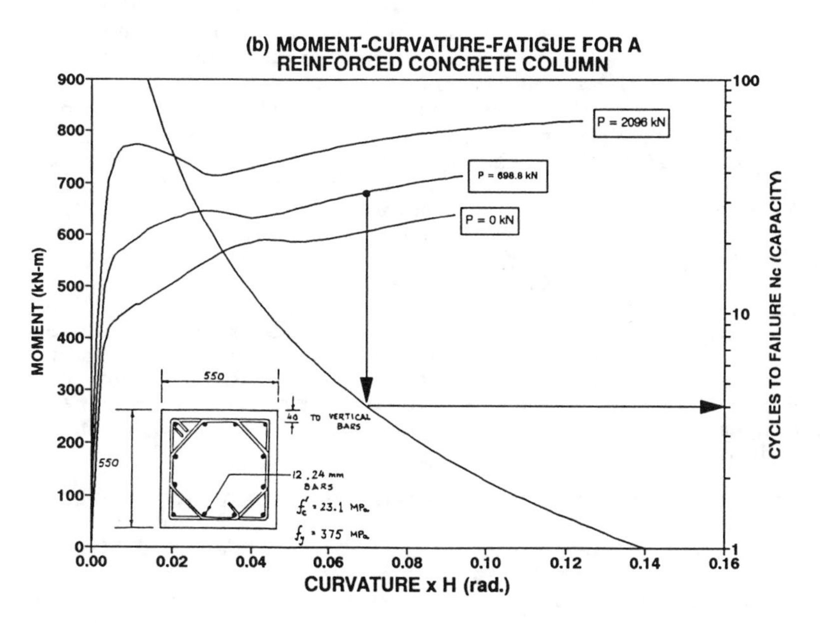


FIG. 3: Hysteretic Capacity of Reinforced Concrete Columns.

reinforced to avoid type (b) and (c) failures described above. If it is assumed that the steel strains will alternate from tension ($\epsilon_s^+$) to compression ($\epsilon_s^-$) under equi-amplitude reversed cyclic loading, then the strain amplitude ($\epsilon_a$) can be defined in terms of the section curvature:

$$\phi = \frac{\epsilon_s^+ - \epsilon_s^-}{(d-d')} = \frac{2\epsilon_a}{(d-d')} \tag{7}$$

where $(d-d')$ = the distance between the outer layers of reinforcement. Hence, Eqs. (6) and (7) can be used to define the equi-amplitude cyclic capacity

$$N_c = 0.0128\,[\phi(d-d')]^{-1/0.5} \tag{8}$$

Now supposing such a column was part of a bridge pier, then for an El Centro type earthquake (with demand $N_d \le 4$), and capacity $N_c \ge 4$ from Eq (8) plotted in Fig. 3(b) $\phi H < 0.07$. If the column has a plastic hinge length of $0.5H$ then this translates into a plastic hinge rotation $\theta_p < 0.033$ radians.

For columns with a high axial load intensity ($P > 0.3\, f_c' A_g$) failure modes (b) and (c) will generally prevail. Studies of bridges in the eastern United States have shown that the axial load intensities are typically low ($P < 0.1\, f_c' A_g$) and the piers generally possess only nominal transverse reinforcement. Thus failure will be characterized by loss of shear strength leading to a type (c) failure. For such squat piers, or columns under-reinforced transversely, inelastic shear displacements will govern the response at moderate to high ductility factors. Reliable inelastic modeling techniques for the shear-failure class of bridge pier are not fully developed at this stage. A clear research need is to continue analytical development in this area. In the meantime, an experimental approach is the most viable means for assessing the post-elastic performance of bridge piers in which shear may control the behavior. An example of a pier with an apparent weakness in shear strength is presented in the next section.

## EVALUATING THE INTRINSIC STRENGTH OF EASTERN U.S. BRIDGES

A survey has been carried out by Chen and Mander (1990) of typical bridges in New York State. This survey showed that for short to medium span highway bridges some 70% of the superstructures are concrete slab on steel girder configurations, and 80% of the substructures are reinforced concrete column, bent or wall piers. A number of such bridges have been evaluated in accordance with the ATC-6-2 (1983) recommendations. For an eastern U.S. type earthquake with a 0.19g peak ground acceleration the following observations were made:

(i) Small to medium span bridges are inherently stiff resulting in low natural periods of vibration (frequencies between 3 and 20 Hz were normal). This in turn demands either a high strength capacity or a high ductility capability if there is insufficient strength for the bridge to respond elastically to the required ground motion.

(ii) The calculated capacity/demand ratios show that these structures generally possess adequate system strength, but the results should be treated with some skepticism. This is because in each case the analysis showed that the shear strength of the system was questionable in the inelastic range.

(iii) Most of the older structures provide only one #3 (10mm) perimeter hoop at 12 inch (300mm) spacing, regardless of the column dimensions. Lap splices are also generally found in the potential plastic hinge zone immediately above spread footings or pile caps. These detailing deficiencies will lead to a questionable seismic performance.

This study indicates that even though bridge piers do possess some intrinsic strength to resist earthquakes inelastically, their ductility capacity is unknown at this stage. Thus, in order to assess the ductility capacity of non-seismically designed bridge piers, experiments are currently in progress. Portions of pier-cap to column connections have been recovered during demolitions and tested under quasi-static reversed cyclic loading in the SUNY at Buffalo Seismic Laboratory. Companion 1/4 scale models of the entire piers have also been constructed and similarly tested to failure.

One such example is shown in Fig 4. This specimen had 1.0% longitudinal steel and only nominal hoops (#3@12"crs in the prototype). The gravity axial load on the columns was $P = 0.05\, f_c' A_g$. It is evident from the experimental results that the pier behaved in a ductile fashion despite poor transverse steel detailing. It will be noted that the pier did not attain its nominal flexural strength capacity under forward cycles of loading, but slightly exceeded it on the reverse cycles. This difference is attributed to variations in material strengths between the two columns.

When testing such piers this raises the question; What displacement history should be used? Many laboratory performance tests histories on columns and piers have in the past been ductility based. The main problem with this approach is explicitly defining the yield displacement.

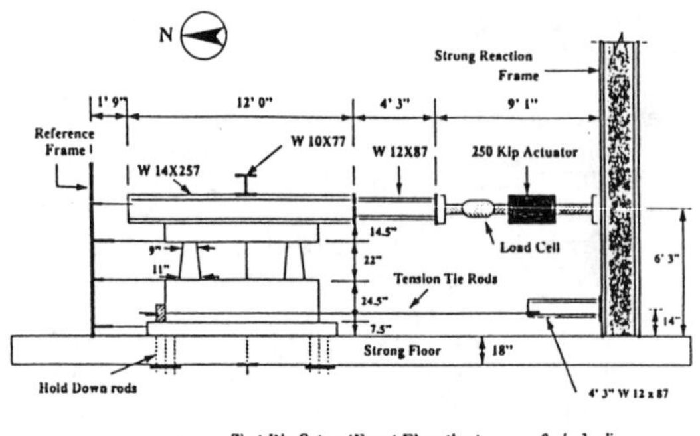


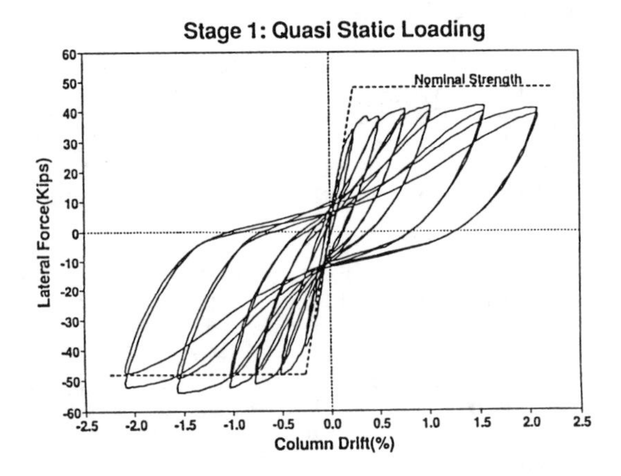


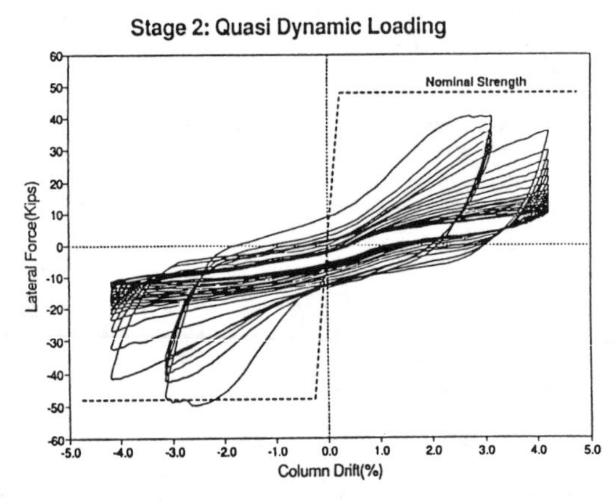


FIG. 4: A One-Quarter Scale Model Bridge Pier Tested Under Cyclic Lateral Loading.

This problem is trivial for flexurally dominated systems. However, for multiple element subassemblages (such as the entire piers shown in Fig. 1) or when the nominal strength cannot be reliably defined, then it is not practical to use ductility based loading histories. Instead, drift based displacement histories can be adopted as shown in Fig. 4 with the number of cycles and/or cyclic amplitude determined from Demand Spectra such as Fig. 2.

**CONCLUSIONS**

Based on the issues raised in this paper the following conclusions are made:

1. The seismic performance of bridges designed in accordance to the code-based practice established following the 1971 San Fernando earthquake should generally be satisfactory.

2. For the type of bridge that exists in low to medium seismic risk zones, existing seismic evaluation techniques are inadequate. An energy-based evaluation methodology described in this paper shows great promise.

3. Non-seismically designed bridges may possess some intrinsic strength that can be utilized for earthquake resistance. However, an adequate seismic performance will depend largely on the reinforcing steel detailing. Squat piers and columns with low volumes of transverse reinforcement may fail in shear. Further research work is needed to determine the energy capacity of this class of pier.

**ACKNOWLEDGEMENTS**

The writers wish to acknowledge the help of graduate student G. A. Chang for help with preparing Fig. 2. The financial support of NCEER is gratefully acknowledged.

**APPENDIX 1: REFERENCES**

_____ "Standard Specifications for Highway Bridges," American Association of State Highway and Transportation Officials (AASHTO), 13th Edition, 1983 and Interim Specifications 1984 and 1985.

ATC-6 (1979),"Earthquake Resistance of Highway Bridges," Proc. Workshop ATC-6, Applied Technology Council, January.

ATC-6-2 (1983), "Seismic Retrofitting Guidelines for Highway Bridges," Report ATC-6-2, Applied Technology Council.

ATC-12 (1981), "Comparison of U.S. and N.Z. Seismic Design Practice for Highway Bridges". Proc. Workshop ATC-12, Applied Technology Council.

ATC-12-1 (1985), "Seismic Resistance of Highway Bridges," Proc. of Second Joint U.S.-N.Z. Workshop ATC 12-1-1, Applied Technology Council.

Chai, Y.H., Priestley, M.J.N., and Seible, F., (1991), "Seismic Retrofit of Circular Bridge Columns for Enhanced Flexural Performance," ACI Structural Journal, Sept-Oct V. 88, No. 5, pp 572-584.

Chen, S.S., and Mander, J.B. (1990), "Seismic Classification, Modeling and Testing of Existing Bridge Systems," Proc. Second Workshop on Bridge Engineering Research in Progress, Reno, Nevada, October 29-30, pp 253-256.

Mander, J.B., Priestley, M.J.N., and Park, R.,(1984) "Seismic Design of Bridge Piers," Research Report 84-2, Department of Civil Engineering, University of Canterbury, Christchurch, New Zealand, 442 pp.

20 authors N.Z.N.S.E.E. (1980), "Papers resulting from deliberations of the Societies' discussion group on seismic design of bridges". Bulletin of the N.Z. National Society for Earthquake Engineering, Vol. 13, No. 3, September, pp 226-309.

Panthaki, F.D. (1991), "Low Cycle Fatigue Behavior of High Strength and Ordinary Reinforcing Steels", MS Thesis, Department of Civil Engineering, SUNY at Buffalo.

Priestley, M.J.N., and Park., R. (1987), "Strength and Ductility of Concrete Bridge Columns Under Seismic Loading," ACI Structural Journal, Jan.-Feb., pp 61-76.

Poland, C.D., White, R.N., Malley, J.O., Gergely, P. (1989), "Seismic evaluation of buildings in the eastern and central United States," in Seismic Engineering Research and Practice, ed. C.A. Kircher and A.K. Chopra, ASCE, NY, pp 546-555.

_______Spectra (1990), "Loma Prieta Earthquake Reconnaissance Report," Earthquake Spectra, Supplement to Vol. 6, pp 151-187.

Uang, C-M., and Bertero, V.V. (1990), "Evaluation of Seismic Energy in Structures," Earthquake Engineering and Structural Dynamics, Vol. 19, pp 77-90.

# Transportation Lifeline Losses in Large Eastern Earthquakes

C. Rojahn[1], C. Scawthorn[2], and M. Khater[3]

Abstract. In December 1991 Applied Technology Council (ATC) and its subcontractor, EQE Inc., completed a 2-year Federal Emergency Management Agency-sponsored study to assess the seismic vulnerability and impact of disruption of lifeline systems nationwide. Included in the study were 489,892 km of federal and state highways, 144,785 bridges, 270,611 km of railroads, 17,161 civil and general aviation airports, and 2,177 ports. Direct and indirect economic losses for 8 scenario earthquakes were estimated using this inventory information and lifeline vulnerability/restoration functions developed under the project. Combined direct and indirect economic losses resulting from damage to transportation lifelines in four Eastern U.S. scenario earth-quakes were estimated as follows: Cape Ann magnitude-7 event, $2,402 million; Charleston, South Carolina magnitude-7.5 event, $6,953 million; New Madrid magnitude-8 event, $14,094 million; and New Madrid magnitude-7 event, $4,378 million.

Introduction. In December 1991 Applied Technology Council (ATC) and its subcontractor, EQE Inc., completed a 2-year Federal Emergency Management Agency-sponsored study to assess the seismic vulnerability and impact of disruption of lifeline systems nationwide. The primary purpose of the project, known as ATC-25, was to develop a better understanding of the impact of disruption of lifelines from earthquakes and to assist in the identification and prioritization of hazard mitigation measures and policies. The results from the investigation are summarized in the ATC-25 report, *Seismic Vulnerability and Impact of Disruption of Lifelines in the Conterminous United States* (ATC, 1991).

Four basic steps were followed to estimate lifeline damage and subsequent economic disruption for eight earthquake scenarios.

1. Development of a national lifeline inventory database that included electric systems; water, gas, and oil pipelines; highways and bridges; airports; railroads; ports; and emergency service facilities.

[1]Executive Director, Applied Technology Council, 555 Twin Dolphin Drive, Suite 270, Redwood City, Calif. 94065

[2]Vice President, EQE International, 44 Montgomery Street, 32nd Floor, San Francisco, Calif. 94104.

[3]Principal Research Engineer, EQE International, 44 Montgomery Street, 32nd Floor, San Francisco, Calif. 94104.

2. Development of seismic vulnerability functions for each lifeline component/system,

3. Characterization and quantification of the seismic hazard nationwide, and

4. Development of direct damage estimates and indirect economic loss estimates for each scenario earthquake.

Reported on in this paper is an overview of the national transportation lifelines inventory and direct and indirect economic losses resulting from damage to transportation lifelines for four scenario earthquakes--magnitude-7 and magnitude-8 events in the New Madrid, Missouri region, a magnitude-7.5 event in Charleston, South Carolina, and a magnitude-7 event in Cape Ann, Massachusetts.

National Transportation Lifeline Inventory

Development of the ATC-25 inventory, for all major lifelines in the United States, was a major task. The project scope required that lifelines be inventoried in sufficient detail for conducting lifeline seismic vulnerability assessments and impact of disruption at the national level. This in turn required that the inventory be compiled electronically in digital form and dictated that inclusion of lifelines at the transmission level, as defined below, was of primary importance.

Initially, a number of government, utility, trade and professional organizations, and individuals were contacted in an effort to identify nationwide databases, especially electronic databases. In most cases, these organizations or individuals referred the project back to FEMA, since they had either previously furnished the information to FEMA, or knew that the data had been furnished to FEMA by others. As a result, FEMA's database became a major source of data for several of the lifelines, including the transportation lifelines.

The transportation lifeline data, which generally date from about 1966 unless later updated by FEMA, consist of networks at the bulk and/or regional level, as opposed to lifelines at the user-level (i.e., distribution level). In other words, the inventory contains only the national *arterial* level, and neglects the distribution or *capillary* system. For example, all federal and state highways are inventoried (Figure 1), but county and local roads are not. The major reason for focusing on the transmission level is that at lower levels the systems only support local facilities. Thus, a disruption of a local activity could not be used to identify the overall regional importance of the lifeline. However, disruptions at the transmission level impact large regions and are therefore important for understanding the seismic vulnerability and importance of lifelines to the United States.

Inventory Overview. The inventory data have been compiled into an electronic database, which generally consists of (i) digitized location and type of facility for single-site lifeline facilities, and (ii) digitized right-of-way, and very limited information on facility attributes for network lifelines. The inventory is only a partial inventory, in that important information on a number of facility

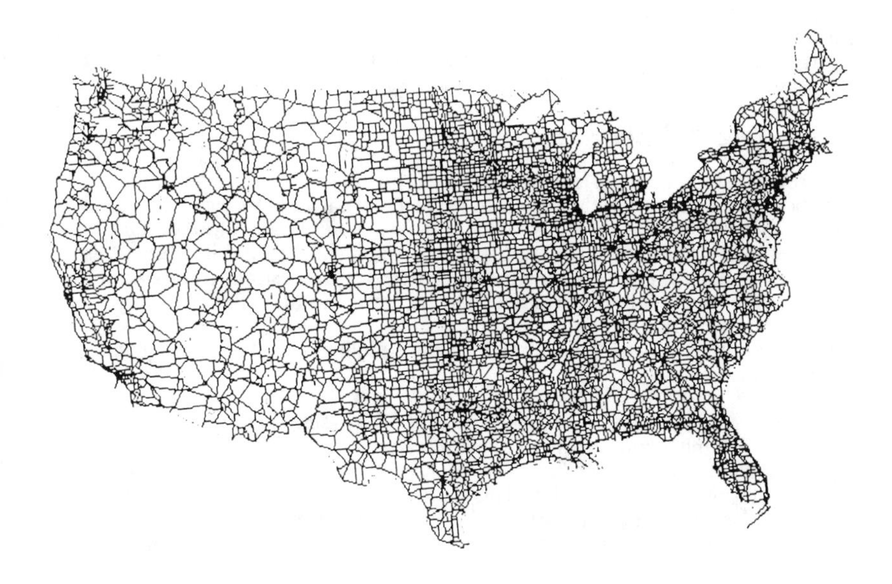

Figure 1. State and Federal Highways

attributes (e.g., number or length of spans for highway bridges) was unavailable from FEMA.

The inventory data include information for the conterminous United States only. Lifeline data for Alaska, Hawaii, and U. S. territories, such as Puerto Rico, have been excluded because lifelines in these regions would not be affected by the scenario earthquakes considered in this study.

The specific transportation lifelines that have been inventoried for the conterminous United States are:

Highways (489,892 km of highway (Figure 1); 144,785 bridges)

Railroads (270,611 km of right-of-way)

Airports (17,161 civil and general aviation airports)

Ports (2,177 ports)

Excluded from the inventory, but included in the analysis, are distribution systems at the local level. For these facility types, the number of facilities in each 25-km by 25-km grid cell, which is the grid size for the seismic hazard analysis, was estimated on the basis of proxy by population.

PC-Compatible Electronic Database. Because the data could also serve as a valuable framework (or starting point) for researchers who wish to investigate lifelines at the regional or local level, including applications unrelated to seismic risk, the data have been formatted for use on IBM-PC compatible microcomputers. The data are unrestricted and will be made available by ATC on 1.2-megabyte, floppy diskettes, together with a simple executable computer program for reading and displaying the maps on a computer screen.

## Lifeline Vulnerability Functions

The second step in the project was the development of lifeline vulnerability functions, which describe the expected or assumed earthquake performance characteristics of each lifeline as well as the time required to restore damaged facilities to their pre-earthquake capacity, or usability. Vulnerability functions were developed for each lifeline inventoried, for lifelines estimated by proxy, and for other important lifelines not available for inclusion in the inventory.

The vulnerability functions developed for each lifeline consist of the following components:

- *General* information, which consists of (1) a *description* of the structure and its main components, (2) *typical seismic damage* in qualitative terms, and (3) *seismically resistant design* characteristics for the facility and its components in particular. This information has been included to define the assumed characteristics and expected performance of each facility and to make the functions more widely applicable (i.e., applicable for other investigations by other researchers).

- Direct damage information, which consists of (1) a description of its basis in terms of structure type and quality of construction (degree of seismic resistance), (2) default estimates of the quality of construction for present conditions and corresponding *motion-damage curves*, (3) default estimates of the quality of construction for upgraded conditions, and (4) *restoration curves*.

These functions reflect the general consensus among practicing structural engineers that, with few exceptions, only California and portions of Alaska and the Puget Sound region have had seismic requirements incorporated into the design of local facilities for any significant period of time. For all other areas of the United States, present facilities are assumed to have seismic resistance less than or equal to (depending on the specific facility) that of equivalent facilities in California NEHRP Map Area 7 (Figure 2). Three regions, representing these differences in seismic design practices, are defined for the United States:

a. California NEHRP Map Area 7, which is assumed to be the only region of the United States with a significant history of lifeline seismic design for great earthquakes,

b. California NEHRP Map Areas 3-6, Non-California Map Area 7 (parts of Alaska, Nevada, Idaho, Montana, and Wyoming), and Puget Sound NEHRP Map Area 5, which is assumed to be the only regions of the United States with a significant history of lifeline seismic design for major (as opposed to great) earthquakes, and

c. All other parts of the United States, which are assumed to have not had a significant history of lifeline seismic design for major earthquakes.

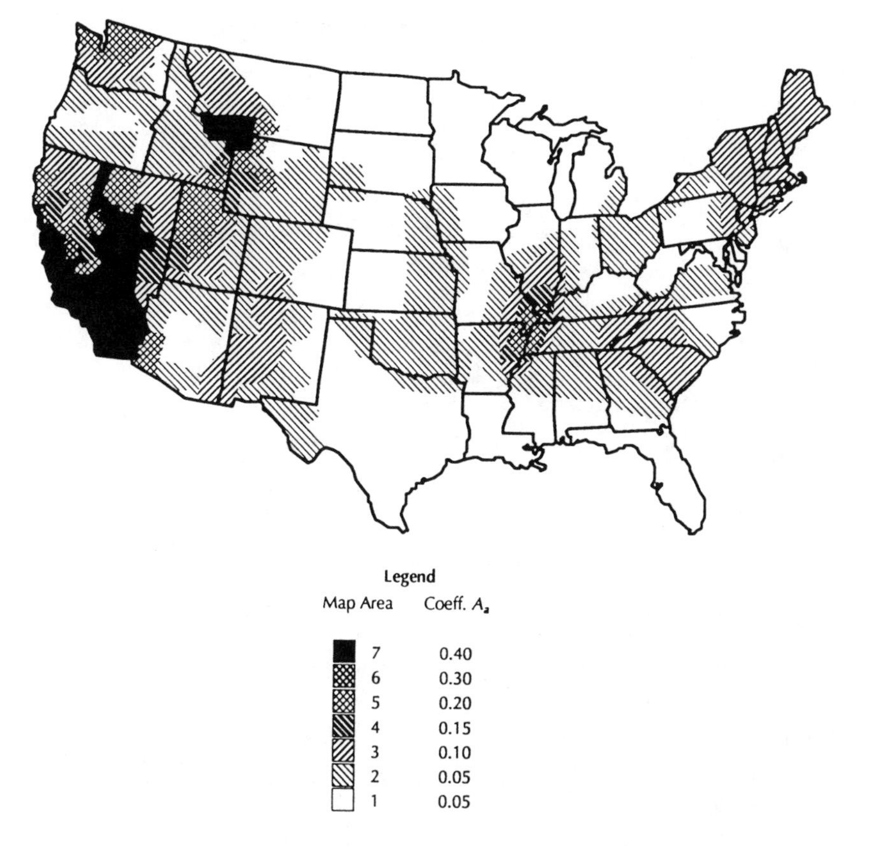


Figure 2. NEHRP Seismic Map Areas (ATC, 1978; BSSC, 1988).

The two key quantitative vulnerability-function relationships developed under this project--*motion-damage curves* and *restoration curves*--define expected lifeline performance for each of these regions and form the heart of the quantitative vulnerability analysis. The curves are based on the data and methodology developed on the basis of expert opinion in the ATC-13 project (*Earthquake Damage Evaluation Data for California*, ATC 1985). Because the ATC-13 data and methodology are applicable for California structures only, however, the data were revised and reformatted to reflect differences in seismic design and construction practices nationwide and to meet the technical needs of the project. All assumptions operative in ATC-13, such as unlimited resources for repair and restoration, also apply to these results.

The *motion-damage curves* developed under this project define estimated lifeline direct damage as a function of seismic intensity (in this case, Modified Mercalli Intensity); direct damage is estimated in terms of repair costs expressed as a fraction or percentage of value. Curves are provided for each

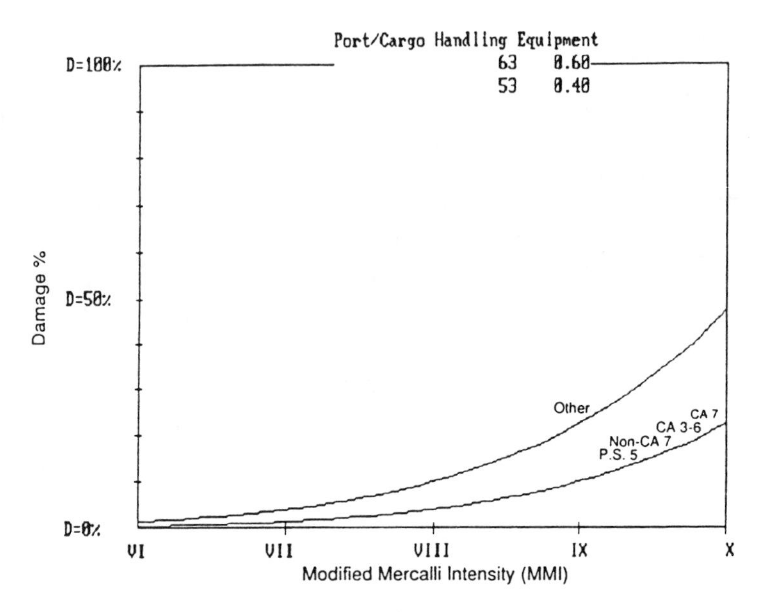


Figure 3. Damage Percent by Intensity for Ports/Cargo Handling Equipment.

region defined above. An example set of motion-damage curves for ports/cargo handling equipment is provided in Figure 3.

The *restoration curves* developed for this project define the fraction of initial capacity of the lifeline (restored or remaining) as a function of elapsed time since the earthquake. Again curves are defined for each region. A sample set is provided in Figures 4 and 5.

## Seismic Hazard

The technical approach for evaluating the seismic hazard of lifeline structures in this project involved identifying (1) the most appropriate means (parameter(s)) for describing the seismic hazard, (2) regions of high seismic activity, (3) representative potentially damaging, or catastrophic, earthquakes within each of these regions that could be used as scenario events for the investigation of lifeline loss estimation and disruption, and (4) a model for estimating the seismic hazard for each of these scenario events.

Descriptor of Seismic Hazard for this Study. Following a review of available parameters for characterizing seismic hazard, we elected to use the Modified Mercalli Intensity (MMI) Scale, a commonly used measure of seismic intensity (effects at a particular location or site). The scale consists of 12 categories of ground motion intensity, from I (not felt, except by a few people) to XII (total damage). Structural damage generally is initiated at about MMI VI for poor

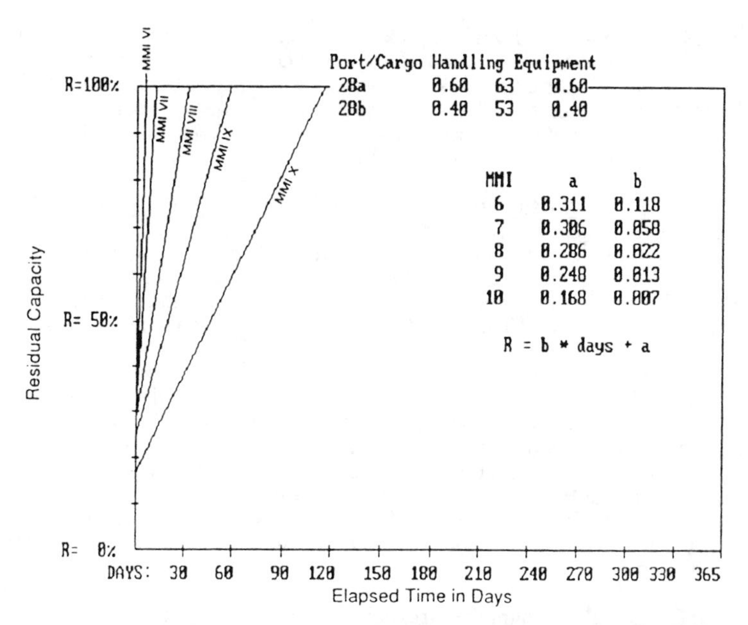


Figure 4. Residual Capacity for Ports/Cargo Handling Equipment (NEHRP Map Area: California 3-5, California 7, Non-California 7, and Puget Sound 5).

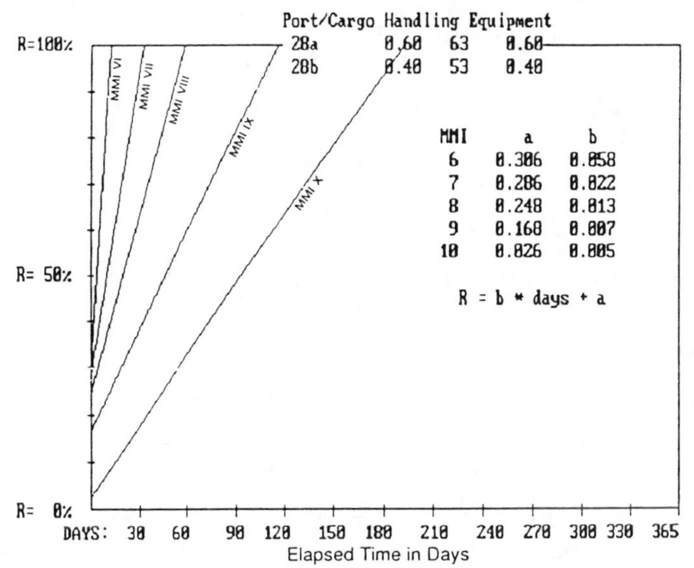


Figure 5. Residual Capacity for Ports/Cargo Handling Equipment (all other areas).

structures, and about MMI VIII for good structures. MMI XI and XII are extremely rare. The MMI scale is subjective; it is dependent on personal interpretations and is affected, to some extent, by the quality of construction in the affected area. Even though it has these limitations, it is still useful as a general description of damage, especially at the regional level, and for this reason was used in this study as the descriptor of seismic hazard.

Choice of a Model for Estimating the Distribution and Intensity of Shaking for Scenario Earthquakes. In order to estimate the seismic hazard (i.e., deterministic intensity) of the scenario events over the affected area associated with each event, a model of earthquake magnitude, attenuation, and local site effects is required. For the conterminous United States, two general models were considered: Evernden and Thomson (1985), and Algermissen et al. (1990).

Selection of one model over the other was difficult, but the Evernden model offered the following advantages for this study: (i) verification via comparison with historical events, (ii) incorporation of local soil effects and ready availability of a nationwide geologic database, and (iii) ready availability of closed-form attenuation relations. An important additional attribute for this project was that the Everden model would estimate the distribution and intensity of seismic shaking in terms of MMI, the shaking characterization used in the ATC-13 study and the basic parameter for the ATC-25 lifeline vulnerability functions.

Scenario Earthquakes. Eight scenario earthquakes were selected for this investigation. In addition to the four events considered in this paper (see introduction for list), the scenario earthquakes investigated in the ATC-25 project included a magnitude-7.5 event along the Wasatch Front (Utah), a magnitude-8 event in Southern California (near Fort Tejon), and a magnitude-7.5 on the Hayward fault in the San Francisco Bay area.

The Evernden model was employed to generate expected seismic intensity distribution in the conterminous United States for the eight scenario events. Shown in Figure 6 is an example intensity distribution for the New Madrid magnitude-8.0 scenario event.

Estimates of Direct Damage

The analysis of seismic vulnerability of lifeline systems and the economic impact of disruption is based on an assessment of three factors:

- Seismic hazard,
- Lifeline inventory, and
- Vulnerability functions.

In this investigation these factors are used to quantify vulnerability and impact of disruption in terms of (1) direct damage and (2) economic losses resulting from direct damage and loss of function of damaged facilities.

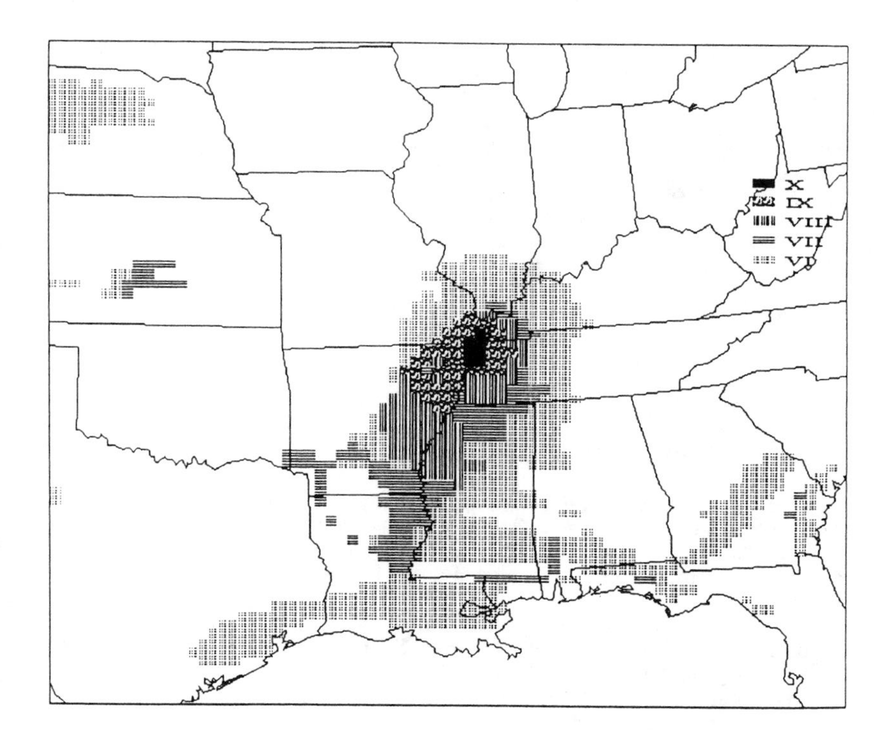


Figure 6. Predicted Intensity Map for New Madrid (Magnitude 8).

Direct damage is defined as damage resulting directly from ground shaking or other collateral loss causes such as liquefaction. For each facility, it is expressed in terms of cost of repair divided by replacement cost and varies from 0 to 1.0 (0% to 100%). In this project it is estimated using (1) estimates of ground shaking intensity provided by the seismic hazard model, (2) inventory data specifying the location and type of facilities affected, and (3) vulnerability functions that relate seismic intensity and site conditions to expected damage.

The analysis approach to estimate direct damage considers both damage resulting from ground shaking as well as damage resulting from liquefaction. Damage due to other collateral loss causes, such as landslide and fire following earthquake, are not included because of the unavailability of inventory information and the lack of available models for estimating these losses nationwide.

The analysis approach for computing direct damage due to ground shaking proceeded as follows. For each earthquake scenario, MMI levels were assigned to each 25-km grid cell in the affected region, using the Everden MMI model, assigned magnitude, and assigned fault rupture location. Damage states were then estimated for each affected lifeline component in each grid cell, using the motion-damage curves developed in this project. The procedure for utilizing the motion-damage curves varied slightly by facility type, depending on whether the lifeline was a site specific facility, or a regional transmission (extended) network.

Site-Specific Lifelines. Direct damage to site-specific lifelines, i.e., lifelines that consist of individual sited or point facilities (e.g., airports, ports and harbors) were estimated using the methodology specified above.

For summary and comparative purposes, four damage states are considered in this study:

- Light damage (1-10% replacement value);
- Moderate damage (10-30% replacement value);
- Heavy damage (30-60% replacement value); and
- Major to destroyed (60-100% replacement value).

Following is a discussion of the direct damage impact on an example transportation lifeline--ports and harbors. Since ports and harbors are located in the coastal regions, only those scenario earthquakes affecting these regions will negatively impact this facility type. For the four eastern United States scenario earthquakes considered in this paper, the most severe damages to ports and harbors are expected for the Charleston event. For example, one hundred percent, or 20 ports and harbors, in South Carolina can be expected to sustain heavy damage (30 to 60%), and 73%, or approximately 22 such facilities would be similarly affected in Georgia. The primary cause of such damage, of course, is poor ground.

Extended Lifeline Networks. With the exception of pipeline systems, which are not considered in this paper, direct damage to extended network lifelines, such as highways and railroads, was also estimated using the methodology specified above. Results are presented in terms of (1) the same four damage states used for site-specific lifelines, and (2) maps indicating the damaged portions of each extended network for the various scenario earthquakes. Example results for an extended lifeline network (railroads) follows.

The railroad system is a highly redundant system, and damage to the system due to the selected events was found to be relatively localized to the epicentral area. Direct damage estimates for the railroad system are based on damage curves for track/roadbed and exclude damage to related facility types not included in the project inventory--railway terminals, railway bridges and tunnels.

The direct damage data suggest that the magnitude-8 New Madrid event would cause the most extensive damage, with 2,265 km of roadbed sustaining damage in the 30 to 100% range. Damage in the Charleston, and magnitude-7.0 New Madrid events would also be severe, with 980 and 640 km of roadbed, respectively, sustaining heavy damage (30-to-60 %). A map showing the distribution of damage to the railroad system for the magnitude-8 New Madrid earthquake scenario is shown in Figure 7.

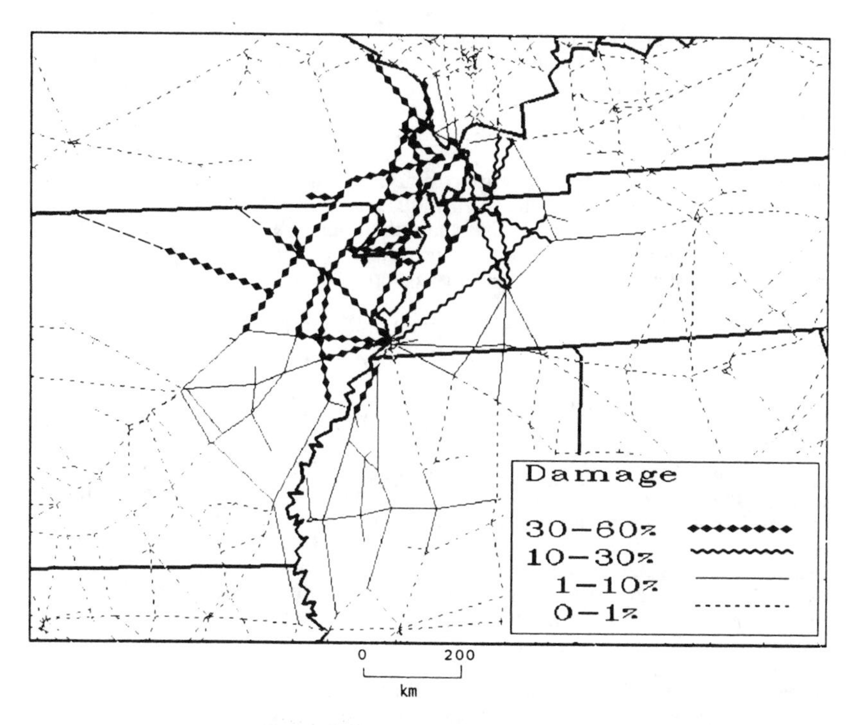


Figure 7. Damage to Railroad System Following Magnitude-8 New Madrid Event.

Dollar Loss Estimates. Summaries of dollar loss estimates for direct damage to regional and local distribution transportation lifeline systems for the four eastern United States scenario earthquakes are provided in Table 1. These data indicate that the most severe direct damages to transportation lifelines occur for the New Madrid magnitude-8 event ($4.485 billion).

**Table 1 Direct Damage Losses to Transportation Lifelines ($ Millions)**

| *Scenario Earthquake* | *Airports* | *Ports* | *Railroads* | *Regional Highways* | *Local Highways* | *Total* |
|---|---|---|---|---|---|---|
| Cape Ann | $91 | $53 | $9 | $382 | $600 | $1,135 |
| Charleston | 142 | 380 | 156 | 773 | 500 | 1,951 |
| New Madrid (M=8) | 411 | 0 | 458 | 2,216 | 1,400 | 4,485 |
| New Madrid (M=7) | 145 | 0 | 108 | 204 | 440 | 897 |

Estimation of Indirect Economic Effects

Earthquakes produce both direct and indirect economic effects. The direct effects, such as dollar loss due to fires and collapsed structures, are obvious and dramatic. However, the indirect effects that these disruptions have on the ability of otherwise undamaged enterprises to conduct business may be quite significant. Although the concept of seismic disturbances and their effect on lifelines has been investigated for at least two decades, there is very little literature on indirect economic losses.

The ATC-25 study provides a first approximation of the indirect economic effects of lifeline interruption due to earthquakes. To accomplish this the relevant literature was surveyed. Then a methodology was developed to relate lifeline interruption estimates to economic effects of lifeline interruption in each economic sector. This required a two-step process:

1. Development of estimates of interruption of lifelines as a result of direct damage

2. Development of estimates of economic loss as a result of lifeline interruption

Estimates of Lifeline Interruption. Lifeline interruption resulting from direct damage is quantified in the ATC-25 investigation in residual capacity plots that define percent of function restored as a function of time. The curves are estimated for each lifeline type and scenario earthquake using (1) the time-to-restoration curves developed from the ATC-13 restoration time data (ATC, 1985), (2) estimates of ground shaking intensity provided by the seismic hazard model, and (3) inventory data specifying the location and type of facilities affected.

For site-specific systems, such as airports, the time-to-restoration curves are used directly whereas for extended regional networks, special analysis procedures are used. These procedures consist of:

- connectivity analyses, and

- serviceability analyses.

Connectivity analyses measure post-earthquake completeness, "connectedness," or "cut-ness" of links and nodes in a network. Connectivity analyses ignore system capacities and seek only to determine whether, or with what probability, a path remains operational between given sources and given destinations.

Serviceability analyses seek an additional valuable item of information: If a path or paths connect selected nodes following an earthquake, what is the remaining, or residual, capacity between these nodes? The residual capacity is found mathematically by convolving lifeline element capacities with lifeline completeness.

A complete serviceability analysis of the nation's various lifeline systems, incorporating earthquake effects, was beyond the scope of the ATC-25 project. Additionally, capacity information was generally not available for a number of the lifelines (e.g, for the highway system, routes were available, but not number of lanes). Rather, a limited serviceability analysis has been performed, based on a set of simplifying assumptions.

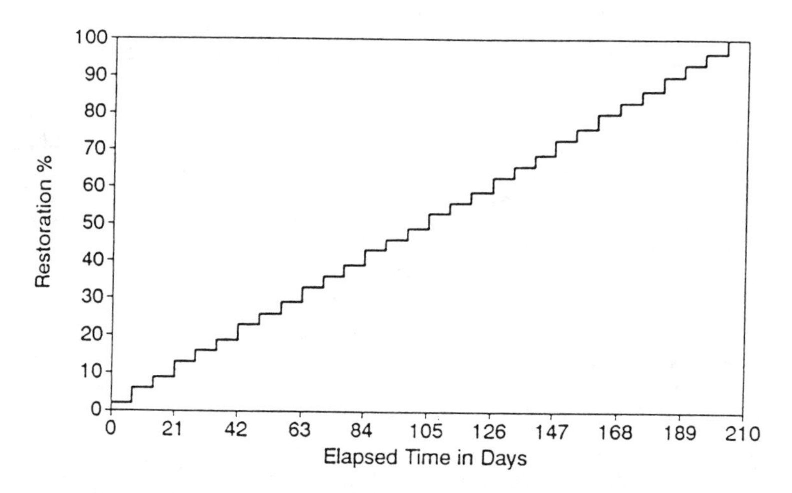


Figure 8. Residual Capacity of South Carolina Ports Following Charleston Event (M=7.5).

The fundamental assumption has been that, on average, all links and nodes of a lifeline have equal capacities, *so that residual capacity has been determined as the ratio of the number of serviceable (i.e., surviving) links and nodes to the original number of serviceable links and nodes, for a given source/destination pair, or across some appropriate boundary.* For example, if the state of South Carolina has 100 airports, and 30 of these are determined to be unserviceable at some point in time following a major earthquake, then the air transport lifeline residual capacity is determined to be 70% of the initial capacity.

An example residual capacity plot for ports damaged in the South Carolina scenario event is provided in Figure 8. In this example, the initial loss is nearly 100 percent of capacity, and full capacity is not restored until about day 200. Georgia would also experience similarly high losses due to the Charleston event. Massachusetts and Rhode Island would experience the largest losses due to the Cape Ann event.

Estimates of Indirect Economic Losses. Indirect economic losses were estimated for each lifeline system and scenario event using the residual capacity plots described above and economic tables showing the estimated percent value-added lost resulting from increasingly severe interruptions to the various lifelines (Value added refers to the value of shipments (products) less the cost of materials, supplies, contract work and fuels used in the manufacture or cultivation of the product). Numerous assumptions, as described in the ATC-25 report (ATC, 1991), were made to estimate the percent loss Value-added data, which are based on the latest available Value-added data (BEA, 1989).

Summaries of the indirect economic losses resulting from damage to the various regional and local transportation lifeline systems, based on 1986 Gross National Product (GNP) data, are provided in Table 2. As indicated in this table, the largest indirect losses for the four eastern United States scenario earthquakes

**Table 2 Indirect Losses Due to Damage to Transportation Lifelines ($ Millions)**

| *Scenario Earthquake* | *Airports* | *Ports* | *Railroads* | *Regional Highways* | *Local Highways* | *Combined Estimate** |
|---|---|---|---|---|---|---|
| Cape Ann | $490 | $450 | $20 | $650 | $860 | $1,267 |
| Charleston | 450 | 4,920 | 20 | 330 | 710 | 5,002 |
| New Madrid (M=8) | 810 | 0 | 250 | 9,360 | 2,000 | 9,609 |
| New Madrid (M=7) | 160 | 0 | 40 | 3,420 | 630 | 3,481 |

*Square root of sum of squares of indirect losses for each lifeline system.

considered herein result from damage to ports (Charleston event) and regional highways (both New Madrid events).

Combined Economic Losses, Deaths and Injuries

Human Death and Injury. It is generally felt that lifeline performance and continuity of operation is vital to human survival in the modern, urban, world. Most observers believe that damage to lifelines would result in human death and injury. Analogous to direct damage to property and indirect economic losses, human death and injury resulting from lifeline damage can be categorized as follows:

1. Human death and injury caused by lifeline functional curtailment, where persons suffer as a result of deprivation of vital services; and

2. Human death and injury resulting from direct damage to lifelines (e.g., occupant injuries resulting from the collapse of an air terminal building).

Casualties Due to Lifeline Functional Curtailment. Without the benefit of hard data it is difficult to estimate with high confidence the number of casualties that will result from curtailment of lifeline function. Our preliminary assessment is that human death and injury due to functional curtailment of lifelines can generally be expected to be very low. Each lifeline was considered, and this conclusion was found to hold, based on the following assumptions: (1) most vital installations that normally require a lifeline service have back-up emergency supplies, and (2) most lifelines have considerable elasticity in demand, and the level of service necessary for life maintenance is very low. In the case of transportation lifelines, which are highly redundant and thus very elastic, emergency food and medicines would be expected to be deliverable regardless of earthquake damage.

Casualties Resulting From Lifeline Direct Damage. Casualties can result from direct damage, especially catastrophic collapse, of lifeline components. Although few deaths occurred directly as a result of lifeline damage in U. S. earthquakes prior to 1989, life-loss due to lifeline failure was tragically demonstrated during the October 17, 1989, Loma Prieta, California, earthquake. Approximately two thirds of the 62 deaths from this earthquake resulted from the failure of a lifeline component--partial collapse of the Cypress structure, a double-decked highway viaduct in Oakland approximately 100 km from the earthquake source zone.

**Table 3 Total Direct Plus Indirect Dollars Losses Resulting from Damage to Transportation Lifelines in Eastern United States Scenario Earthquakes ($ Millions)**

| *Scenario Earthquake* | *Airports* | *Ports* | *Railroads* | *Regional Highways* | *Local Highways* | *Combined Estimate** |
|---|---|---|---|---|---|---|
| Cape Ann | $581 | $503 | $29 | $1,032 | $1,460 | $2,402 |
| Charleston | 592 | 5,300 | 176 | 1,103 | 1,210 | 6,953 |
| New Madrid (M=8) | 1,211 | 0 | 708 | 11,576 | 3,400 | 14,094 |
| New Madrid (M=7) | 305 | 0 | 148 | 3,624 | 1,070 | 4,378 |

*Sum of Direct Losses for each Lifeline plus Combined Estimate for Indirect Losses (Table 2)

Although it can be argued that the deaths and injuries caused by lifeline failure in the Loma Prieta earthquake were the exception, not the rule, the vulnerability functions developed for this project suggest that substantial life-loss from lifeline component failure should be anticipated. Lifeline failures that could cause substantial life loss or injury include bridge failure and railroad derailment.

Combined Direct and Indirect Economic Losses. Summaries of total direct and indirect dollar losses resulting from damage to transportation lifelines for each eastern United States scenario earthquake are provided in Table 3.

## References

Algermissen, S. T. et al., 1990, "Probabilistic Earthquake Acceleration and Velocity for the United States and Puerto Rico," U. S. Geological Survey Map MF-2120.

Applied Technology Council (ATC 3-06), 1978, *Tentative Provisions for the Development of Seismic Regulations for Buildings*, Report ATC 3-06, Palo Alto, Calif.

Applied Technology Council (ATC-13), 1985, *Earthquake Damage Evaluation Data for California*, Report ATC-13, Redwood City, Calif.

Applied Technology Council (ATC-25), 1991, *Seismic Vulnerability and Impact of Disruption of Lifelines in the Conterminous United States*, Report ATC-25, Redwood City, Calif.

BEA, February 1989, "Survey of Current Business," Vol. 69, No. 2, pp. 21-36, Input-Output Accounts of the U.S. Economy, 1983.

BSSC, 1988, *NEHRP Recommended Provisions for the Development of Seismic Regulations for New Buildings*, Federal Emergency Management Agency, Eaerthquake Hazard Reduction Series Report 95, Washington, DC.

Evernden, J. F., and J. M. Thomson, 1985, "Predicting Seismic Intensities, in Evaluating Hazards in the Los Angeles Region--An Earth Science Perspective," J. I. Ziony, ed., U. S. Geological Survey Prof. Paper 1360, U. S. GPO, Washington, DC.

# Regional Evaluation of Transportation Lifelines in New York State with the Aid of GIS Technology

Masanobu Shinozuka [1], M. ASCE, Michael P. Gaus [2], M. ASCE,
Seong H. Kim [3] ,SM, ASCE, and George C. Lee [4], M. ASCE

## ABSTRACT

Regional evaluation of transportation lifeline facilities which may be impacted by seismic and other natural hazards is of great concern for purposes of emergency planning and response and for establishing priorities for maintenance and retrofit if an unacceptable risk is found to exist. An important tool which can aid in carrying out such an analysis is the use of computer-based Geographic Information Systems (GIS). Use of a GIS system to perform interactive regional risk analysis for bridges is described to illustrate the procedures involved. The analysis utilizes the GIS system as a platform to integrate the wide variety of information and computational procedures needed to evaluate the impact of a user selected earthquake on the total system of bridges in a region. Erie County, NY is utilized as a demonstration area for regional analysis which is representative of conditions existing in the eastern U.S.

## 1. INTRODUCTION

The evaluation of the seismic vulnerability of bridges and other lifeline facilities is a problem of current vital interest. Evaluation may consider individual structures for the purpose of reaching decisions regarding needed retrofit or upgrade procedures or may be carried out with a systems point of view in which the potential impact of a seismic event would be evaluated for all facilities in a regional area. A regional evaluation could provide a valuable interactive tool for emergency management and response procedures. Studies have been carried out to estimate the seismic vulnerability of lifelines on a national basis[2], for specific systems such as bridges in zones of higher seismic activity[7] and water delivery systems[9].

The focus of this paper is on the regional evaluation of the seismic vulnerability of lifeline systems in zones of lower frequency seismic events such as is the case for New York State which is representative of conditions in the eastern U.S. In these regions the level of information available for seismic evaluation of lifeline facilities has not been as well developed as is the case for zones of more frequent seismic activity. A GIS-based approach is described as a platform to integrate the wide variety of information needed to evaluate the impact of earthquakes or other natural hazards on a

1. Prof., Dept. of Civil Engr. and Director, National Center for Earthquake Engineering Research, SUNY at Buffalo, Buffalo, NY 14260
2. Prof. Dept. of Civil Engineering, SUNY at Buffalo, Buffalo, NY 14260
3. Ph.D. Candidate, Dept. of Civil Engineering, SUNY at Buffalo, Buffalo, NY 14260
4. Prof. and Dean of School of Engr., SUNY at Buffalo, Buffalo, NY 14260

regional network of primary and secondary lifeline facilities. Specifically the case of bridges will be utilized to illustrate the methodology of the assessment procedure. The GIS is used to provide real-time interactive capability, for efficient database management, and as a tool to integrate earth science, structural characteristics, topological system characteristics and other vital information required for a risk based model which can be utilized for decision making.

Ideally it would be desired to utilize as wide a variety of data as possible in carrying out a GIS-based seismic vulnerability analysis so that all relevant parameters are considered in the analysis. At present, this is a difficult task to achieve due to data residing with a large variety of sources and in a wide variety of forms for the information. Even in cases where work has started to encode spatial data and attributes in digital electronic form there are still many problems in formatting the data so that problems of scale, registration, type of geographic representation and coordinate system, resolution, file structure, attribute files and other information are consistent. Progress is being made in dealing with these problems although this progress is not uniform even across a statewide geographic region such as New York State and it will take some time and considerable effort to deal with these problems.

In order to illustrate the procedures involved in utilizing GIS as an integrating tool in regional risk assessment of lifelines subjected to natural hazards, an example will be used involving one county, Erie County, New York and one type of lifeline element, bridges[6]. The analysis is carried out using information currently available in suitable form on a wide geographic basis. A modular approach is utilized with an open system approach so that as additional data becomes available and as improved damage assessment models are developed, source data refined, soil characteristics better identified, attenuation models refined and other information made available, this new information can be added to or substituted for existing portions of the assessment program. In this way the system can be readily improved in the future.

## 2 GENERAL ANALYTICAL APPROACH

The overall approach developed for this study involves three major interactive components:

(a) The use of GIS to provide the interface to display geographic data, to manipulate information stored in relational data bases, to calculate topological parameters such as distance to a selected source, to sequentially call subprograms which evaluate vulnerability conditions for each bridge in the selected analysis region and to display the result of a query for the effect of a selected source and magnitude event.

(b) A risk model for bridges which can statistically predict the expected level of damage due to a particular intensity of ground motion at the bridge site.

(c) A ground motion attenuation model to predict the intensity of ground motion at a particular bridge site due to energy release at a selected potential source.

These interactive components are supported by data files which encode characteristics such as potential earthquake sources and magnitudes, and bridge characteristics which are important for failure analysis. As is usual in a GIS system, these files can be generated externally but in a format compatible with the GIS system.

## 3 GIS DATA BASES

The assembly of geographic data for a particular application requires the consideration of many parameters such as scale and resolution, feature detail, organization of layers for information, coordinate system and others. A particularly convenient source of GIS data are the TIGER data base files which were generated for the 1990 census. For this particular study, a commercially enhanced TIGER data file for Erie County, New York was utilized which was assembled from thirty one 7.5 Minute Quad maps which are available for most urban areas of the U.S. Although spatial resolution of the Tiger files only provide the location of a bridge within 75 feet, it was felt that this was quite adequate for this study, particularly in view of the uncertainties inherent in the earthquake source data and the attenuation models available. Figure 2 shows a portion of Erie County at the 1:24,000 scale of a 7.5 Minute Quad map. This provides an overall view of an area and it is possible to identify major features of the transportation road network and other major features. A great deal of detail is embedded in this map. However, the detail cannot be distinguished at this scale. The GIS system provides a capability to zoom-in or window on a section of map. Figure 3 shows a windowed section of the map in Figure 2. As can be seen in this Figure, details such as street names have now become visible on the enlarged section of the map. Analysis is being carried out using the ARC/INFO system running on SUN workstations.

## 4 DAMAGE AND FAILURE ASSESSMENT MODEL

A general screening model for GIS-based risk analysis of bridges in regional environments was developed for this study. The procedure followed was to first collect as much data as could be readily found in the published literature regarding bridge failures and damage from earthquakes. These data were then reviewed to determine if sufficient detail was included in descriptions to permit an identification of the particular elements of the bridge system which resulted in failure and if sufficient information was provided to arrive at a fairly reliable estimate of ground intensity at the bridge site. Relatively complete data were found for 74 bridges located mainly in California which were damaged or failed during an earthquake. These data were examined to establish a suitable classification scheme which could encompass the major variables related to bridge performance, damage and failure. Some of the features considered are bridge type, types of components which were involved in damage or failure, and characteristics of loading. The last item is related to the potential earthquake source mechanisms and earthquake ground motion attenuation which is discussed in a later section of this paper.

A major problem in extracting damage and failure information from published sources was the lack of any standard for reporting such data. In order to attempt to establish a reasonably consistent procedure for data evaluation, a series of levels of classification were established. On an overall or macro level, the parameters can be grouped into loading environment or site intensity due to a given earthquake, degree of damage, soil conditions, foundation type and structural parameters such as pier details, materials used and details such as bearing type. Clearly bridge failures or damage could be initiated by liquefaction or surface faulting. However, in order to reduce the number of parameters to be considered in this study, these types of failure conditions were deferred for future study and primary concentration was placed on damage or failure due to ground shaking rather than ground failure. The 74 bridges used in this study excluded these types of ground failure.

A series of general (or perhaps fuzzy) categories were selected for the initial work on this problem. These categories are rather broad as the data available is also rather fuzzy and a high degree of computational precision does not seem appropriate.

A listing of the parameters utilized is given in Table 1 based on previous researchers' work and on the authors' own study. One important consideration in selecting parameters is the information which is available for bridges in the area to be studied and displayed using the GIS system.

For the identification and characterization of the bridges in Erie County, New York which was used as the demonstration area, a tape of the New York State bridge inventory and inspection data was obtained through the courtesy of the New York State Department of Transportation. The bridges in Erie County were then extracted from the larger state database along with the desired standard data categories for these bridges. The data selected for damage or failure evaluation from bridges actually subjected to damaging earthquakes were matched to the data available from the NYS inventory.

After data identification and classification was completed, a statistical analysis was performed utilizing a standard multiple regression technique. Four intensities of ground motion were selected using peak acceleration value as the defining criteria. These categories are shown in Table 1. The regression analysis simply evaluates the potential contribution of each parameter to the level of damage for each bridge in the database used. The resulting damage equations $y_j$ have the following form:

$$y_j = \sum_{i=1}^{N} \beta_i \cdot X_i + C$$

where $y_j$ is the damage or failure level as indicated in Table 2. These failure levels were known for each of the 74 bridges in the data set used and N=14 corresponds to the number of primary categorization parameters selected. The system of 74 equations was solved to determine values for the beta's and C. The beta's, X's and C are shown in Table 1. This rather crude analysis provides a rough indication of how each parameter contributes to the statistical level of damage or failure for the bridges for which data were available.

The model developed above represents the damage probability for the entire ensemble of bridges in the data set available. It is also necessary to determine the level of reliability which will result when this model is applied to one individual bridge in the set.

Using the 3 group classification shown in Table 2, the actual and predicted rank of seismic vulnerability of bridges were compared. The results are shown in Table 3. This table shows that the ranks of seismic vulnerability of 57 bridges among the 74 bridge data set are predicted correctly. In other words for this set, there is a 77% probability of correct prediction, which seems to be good enough for this initial GIS-based study. Due to the procedure used, the probability of correct prediction is almost even for bridges in different ranks of seismic damage as shown in Table 4.

Many other damage or failure models could be formulated. For example another approach would be to evaluate the level of damage to each class of bridge for the particular code which was in effect in the year when the bridge was designed or modified. Ultimately it would be desirable to develop a damage or failure model which could be directly evaluated from each actual bridge design, or from field data collected specifically for this purpose and from an assessment of the current "state-of-health" of the particular bridge.

The damage equation above is applied to 33 bridges in Erie County under the assumed earthquake intensity of 0.1g and 0.2g. Under 0.1g intensity all bridges are rated low for their damage vulnerability except for 8 which are rated moderate. In case of 0.2g intensity, 17 bridges are rated low, 12 moderate and 4 high.

## 5 EARTHQUAKE SOURCE DATA AND ATTENUATION MODEL

For a given study area, the locations of possible earthquake sources which release energy and the variation of surface intensity of ground motion with distance from the source are needed. Clearly each of these topics could be and has been a subject of intense study on their own. In fact much information on the attenuation models is available from sources such as the U.S. Geological Survey, the New York State Geological Survey, the Electric Power Institute and the National Center for Earthquake Engineering Research.

An example of the type of information which is available is a recent study supported by the Electric Power Research Institute (EPRI)[5] in which the eastern United States was evaluated with respect to various seismicity parameters and estimates were made of the maximum magnitudes that a given seismic source might generate.

The risk analysis procedure is a general methodology and once developed can easily accommodate different source data or attenuation relationships. Because the data on probable magnitudes is extremely sparse, the GIS-based procedure is set-up to allow a user to carry out "what-if" analysis and to assign various magnitudes to source events to evaluate the potential consequences.

Attenuation relationships have been studied by a number of researchers[3,4,5,8,10]. Problems which must be addressed involve estimating a value for epicentral surface intensity for a particular source and set of geological conditions and then to estimate the attenuation relationship for a given geographic region. The problem is further complicated because surface ground motion intensity is also highly sensitive to local soil or geological conditions and to frequency content of the waves propagated from the source and repropagated from geological discontinuities both below and on the surface. Due to the preliminary nature of this study, however, the bridges in Erie County considered are assumed to be subjected to either 0.1g or 0.2g seismic intensity as mentioned above.

## 6 INTERACTIVE GIS-BASED ANALYSIS

With the basic procedures for risk analysis defined, the integration with the GIS system can be formulated. As the inventory of bridges ranges from Interstate bridges to local bridges, a system of icons to identify the bridges and their importance in the regional transportation network is required. For each bridge, a table or file is prepared containing the attributes which are required for the risk evaluation (i.e. soil conditions, foundations, piers, etc.).

An attribute table is also prepared for possible sources and their data encoded or digitized. Each source can include potential ranges of magnitudes and their probabilities. As an option, the user can specify an arbitrary source and magnitude and display impacts on the bridge network.

Menu bars are defined to display choices such as source locations, bridge types and similar information. For each of these classes, a submenu offers choices such as sources likely to have magnitudes greater than a certain level or between certain bounds.

The overall flow diagram for the program is shown in Figure 1.
Graphic data are stored in layered databases and tabular data are interactive with the graphic data. New information generated is contained in mapoverlays.

Standard features of the ARC/INFO program are used for map file input, zooming, print-out and report generation. The use of the GIS system provides a new dimension for engineering analysis and planning.

## 7 CONCLUSIONS

GIS systems provide a new tool and opportunities for interactive analysis of problems such as regional or spatial risk analysis. The GIS system by itself is only an added tool and does not remove the necessity to formulate appropriate analysis techniques for evaluation of problems such as risk assessment for bridges. A valuable feature of the GIS-based approach is that it can provide a general "open-system" methodology in which components such as the damage model or even type of hazard could be easily modified or substituted without having to remanufacture the entire system. The risk analysis model developed for this study could be additive, as more factors and research findings are gathered and the accuracy of correct expectation of future damage state will be further improved. The role of GIS in integrating these modules into a harmonic interactive system will find increasing use in the future for studying large-scale engineering problems. The societal importance issue was not discussed here but it will be in the future.

## ACKNOWLEDGEMENT

This work was partially supported by NCEER under Award Number 913313.

Table 1. Components of Risk Assessment Model

| VARIABLES | CLASSIFICATION | BETA'S & C |
|---|---|---|
| y = Degree of Damage | o;no 1;minor 2;moderate 3;severe 4;falling-off of super structure | |
| X1 = Intensity of Peak Ground Acceleration | 1; $A<0.1G$ 2; $0.1G \le A<0.2G$ 3; $0.2G \le A<0.3G$ 4; $0.3G \le A$ | 0.222 |
| X2 = Design Specification | 1; before 1940 2; 1940-1971 3; 1972-1980 4; after 1981 | -0.358 |
| X3 = Type of Superstructure | 1; simply supported girder-2 spans or more or 2 level or more elevated model 2; simply supported girder-single span or continuous girder 3; arch, frame, cable-stayed, or suspension bridges | -0.234 |
| X4 = Shape of Superstructure | 1; straight 2; skewed or curved | 0.500 |
| X5 = Material of Superstructure | 1; steel 2; RC or PC 3; timber, masonry or other old materials | -0.098 |
| X6 = Internal Hinge | 1; not exist 2; exist | 0.373 |
| X7 = Type of Pier | 1; solid or frame 2; individual columns-2 or more 3; individual columns-single | -0.065 |
| X8 = Type of Foundation | 1; spread footing 2; footing on piles 3; foundations designed by 1983 specification or later | 0.231 |
| X9 = Material of Substructure | 1; steel 2; concrete 3; timber, masonry or other old materials | 0.514 |
| X10 = Height of pier | 1; $H<15$ Ft 2; $15 \le H<30$ Ft 3; $H \ge 30$ Ft | 0.101 |
| X11 = Irregularity in Geometry or in Stiffness | 1; no 2;yes | 0.562 |
| X12 = Site Condition | 1; type 1 2; type 2 3; type 3 (AASHTO 1988) (ref.1) | 0.421 |
| X13 = Effect of Scouring | 1; none 2; recognized | -0.273 |
| X14 = Seat Length | 1; good 2; fair 3; poor | 0.478 |
| | | C=-2.243 |

Table 2. Rank of Seismic Vulnerability

| Rank of Damage Degree | Rank of Vulnerability |
|---|---|
| 0: no damage 1: minor damage | C: Low |
| 2: moderate damage | B: Moderate |
| 3: severe damage<br>4: falling-off of Superstructure | A: High |

Table 3. Comparison of Actual and Predicted Rank of Seismic Vulnerability

| Rank | | Predicted Vulnerability Rank | | | Total |
|---|---|---|---|---|---|
| | | A | B | C | |
| Actual Vulnerability Rank | A | 14 | 4 | 0 | 18 |
| | B | 3 | 13 | 1 | 17 |
| | C | 1 | 8 | 30 | 39 |
| Total | | 17 | 26 | 31 | 74 |

Table 4. Probability of Correct Prediction

| | |
|---|---|
| A | 78% |
| B | 76% |
| C | 77% |
| Total | 77% |

Figure 1. GIS - Model Interaction Flowdiagram

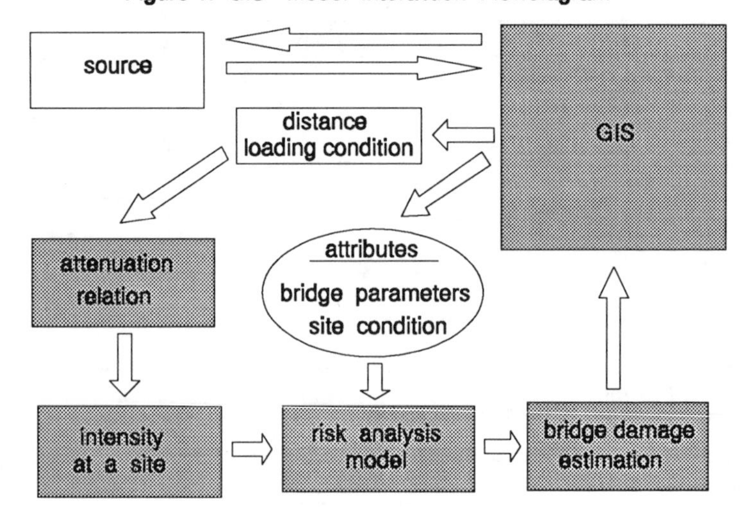

Figure 2. 1:24,000 scale map

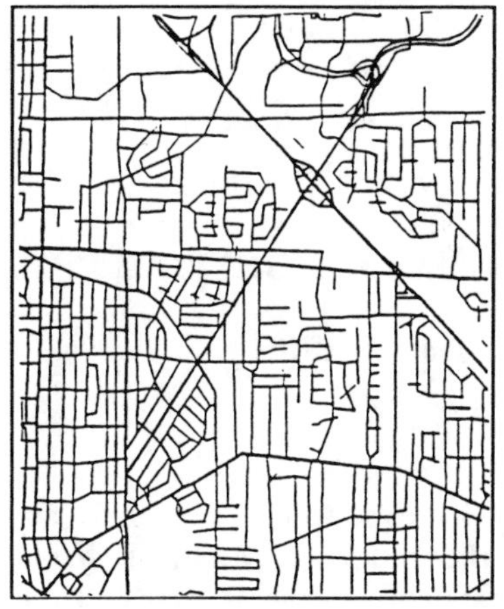

Figure 3. Enlarged Area

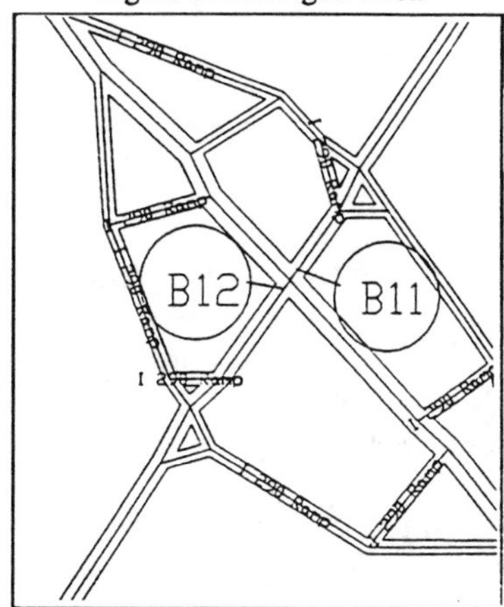


# REFERENCES


1. AASHTO. (1988) 'Guide Specifications for Seismic Design of Highway Bridges 1983' includes Revisions from:Interim Specifications, Bridges 1985; Interim Specifications, Bridges 1987-1988;: Washington,D.C.

2. Applied Technology Council (1991), 'ATC-25, Seismic Vulnerability and Impact of Disruption of Lifelines in the Conterminous United States' Applied Technology Council, Redwood City, CA.

3. Atkinson,Gail M. (1984) 'Attenuation of Strong Ground Motion in Canada From a Random Vibration Approach' BSSA,Vol.68,No.4,pp.1147-1179.

4. Bolt,B.A. (1978) 'Fallacies in Current Ground Motion Prediction' in Proceedings of the 2nd International conference on Microzonation, Vol.II., pp.617-634.

5. Electric Power Research Institute. (1986) 'Seismic Hazard Methodology for the Central and Eastern United States' EERI report NP4726, EERI, Palo Alto, Cal.

6. Kim, S.H., Gaus, M.P., Lee, G.C. and Chang, K.C. (1992) 'GIS-Based Regional Risk Approach for Bridges Subjected to Earthquakes' in Proceedings of the 8th conference on computing in Civil Engineering, ASCE, 1992, pp.460-467.

7. Maroney, Brian and Gates, James, 'Seismic Risk Identification & Prioritization in the CALTRANS Seismic Retrofit Program', Proceedings of the Fourth International Conference on Seismic Zonation, Aug., 1991, Stanford University, Stanford, CA.

8. McGuire,R.K. (1977) 'Effects of Uncertainty in Seismicity on Estimates on Seismic Hazard for the East Coast of the United States' BSSA, Vol.67, No.3, pp.827-848.

9. Sato, R. and Shinozuka, M, 'GIS-Based Interactive and Graphic Computer System to Evaluate Seismic Risks on Water Delivery Networks', in Proceedings of the Fourth International Conference on Seismic Zonation', Aug., 1991, Stanford University, Stanford, CA.

10. Thiel,Jr.C.C., Boissonnade,A.C., Miyasoto,G.H. (1986) 'An Assessment of Eastern United States Strong Ground Motion Attenuation Relationships' in Proceedings of the 8th European Conference on Earthquake Engineering, Vol.1, pp.3.1/63-70.

# SEISMIC HAZARD ALONG A CENTRAL U.S. OIL PIPELINE

Howard H. M. Hwang,[1] Member ASCE

## ABSTRACT

An integrated approach is used to evaluate seismic hazards at six sites where pipeline 22 crosses the major rivers in West Tennessee. On the basis of historical and instrumental data, a recurrence relationship for the New Madrid seismic zone is established. Then, the seismic hazard curves for these six sites are developed by using the ground motion attenuation relationship proposed by Nuttli and Herrmann. In addition, the hazard-consistent magnitude and hazard-consistent distance corresponding to a specified annual probability of exceedance are determined to establish the ground motion such as earthquake acceleration time histories of a probability-based scenario earthquake. These acceleration time histories can then be used for evaluating soil liquefaction potential at a site and seismic vulnerability of lifeline facilities.

## INTRODUCTION

Three large crude oil pipelines are located within or near the New Madrid seismic zone (NMSZ) (Beavers et al. 1986). One of these pipelines is pipeline 22, a 40-inch-diameter pipeline that travels through soft sediments of the Mississippi Valley. The pipeline is not designed to resist earthquakes. If the pipeline were ruptured during an earthquake, the spilled oil could contaminate the recharge area for the water supply in West Tennessee. Such an

---

[1] Professor, Center for Earthquake Research and Information, Memphis State University, Memphis, TN 38152

impact would be enormous since there is no immediate alternative source of water supply in this region.

Pipeline 22 crosses six major rivers in West Tennessee: Wolf River, Loosahatchie River, Hatchie River, Forked Deer-South Fork, Forked Deer-North Fork, and Obion River. These river crossings (Table 1 and Figure 1) are the potential locations of soil liquefaction during a large earthquake and therefore are selected as the sites for seismic hazard evaluation. In this study, seismic hazards are evaluated for these six sites along pipeline 22.

Table 1 Site Locations

| Site | River | Latitude | Longitude |
|---|---|---|---|
| 1 | Wolf River | 35.07° | 89.63° |
| 2 | Loosahatchie River | 35.28° | 89.56° |
| 3 | Hatchie River | 35.55° | 89.44° |
| 4 | Forked Deer-South Fork | 35.80° | 89.36° |
| 5 | Forked Deer-North Fork | 35.96° | 89.32° |
| 6 | Obion River | 36.25° | 89.17° |

## PROBABILISTIC SEISMIC HAZARD ANALYSIS

Seismic hazard at a site depends on the source of earthquakes, seismicity in the vicinity of the site, and attenuation of the ground motion. By performing a probabilistic seismic hazard analysis, the annual probability of exceedance of a seismic intensity parameter, for example, response spectra or peak ground acceleration (PGA), can be evaluated by considering uncertainties in seismic source zones, seismicity, and path attenuation including the local soil effect. The response spectra are more desirable for engineering application; however, attenuation of spectral ordinates is not well established in the NMSZ. Thus, seismic hazard is still be estimated in terms of the PGA and the result is displayed as a seismic hazard curve.

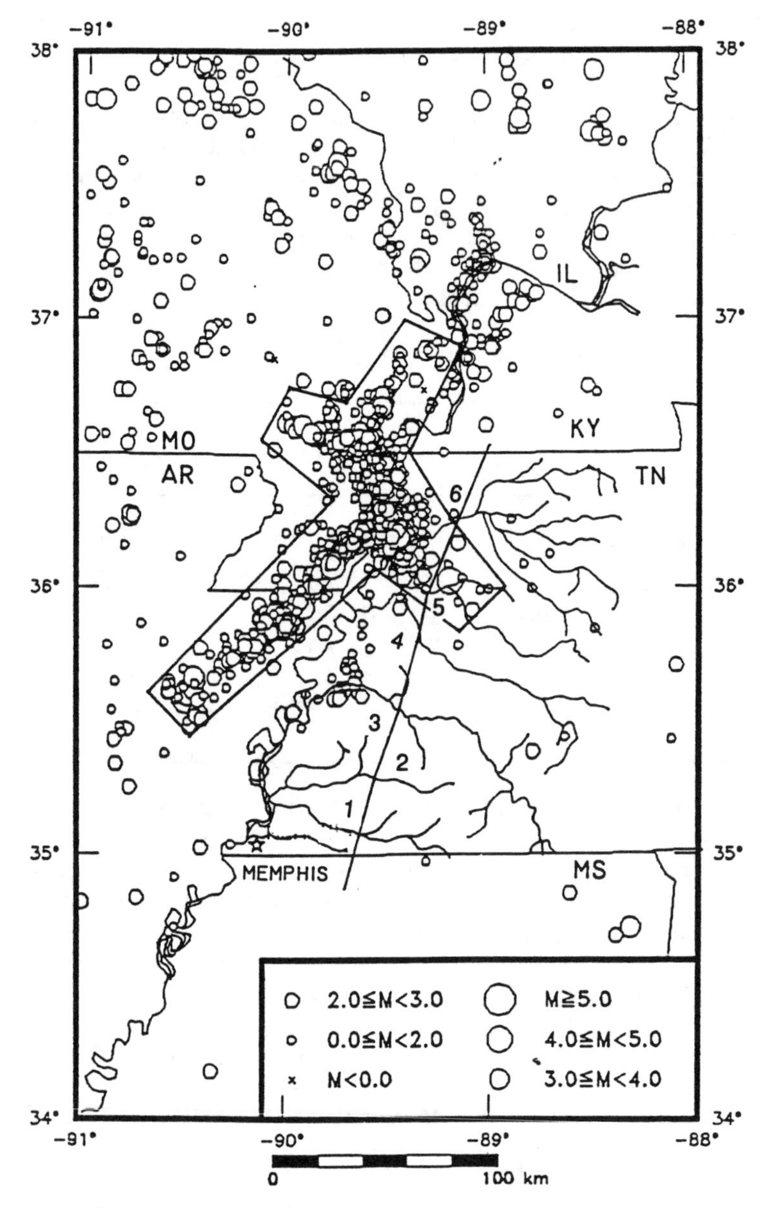


Fig. 1 New Madrid Seismic Zone and Site Locations

## NEW MADRID SEISMIC ZONE

The New Madrid seismic zone, delineated by the concentration of epicenters of earthquakes (Figure 1), consists of three fault segments: (1) a southern segment extending from Marked Tree, Arkansas, to Caruthersville, Missouri, roughly along the axis of the Reelfoot rift complex; (2) a middle segment trending northwest and extending from Ridgely, Tennessee, to west of New Madrid, Missouri; and (3) a relatively shorter northern segment extending from west of New Madrid, Missouri, to southern Illinois. Johnston and Nava (1990) divided the NMSZ into two subzones: zone A and zone B. Because the recurrence rate in zone B is much smaller than that in zone A, the contribution of zone B to the overall seismic hazard at a site is usually small. In this study, only zone A suggested by Johnston and Nava is used to represent the NMSZ (Figure 1). For the case that probabilistic seismic hazard analysis is performed rigorously, however, other seismic zones may need to be included.

## RECURRENCE RELATIONSHIP

A recurrence relationship indicates the chance of an earthquake of a given size occurring anywhere inside the source zone during a specified period of time (one year in this study). The recurrence of earthquakes in a source zone is usually expressed as follows (Gutenberg and Richter 1944):

$$\log N = a - b\, m_b \tag{1}$$

or

$$N(m_b) = e^{\alpha - \beta\, m_b} \tag{2}$$

where a is the log of the number of earthquakes of magnitude zero or greater expected to occur during the specified period of time; b is the slope of the recurrence line; $\alpha = a \ln 10$; $\beta = b \ln 10$; $m_b$ is the body-wave magnitude; and N is the cumulative number of earthquakes that have magnitude of $m_b$ or greater. The data base for the New Madrid seismic zone is a combination of historical data (1804-1974) and instrumental data (1974-1990). Gordon (1988) relocated several earthquakes in the New Madrid zone. Thus, the original data base is modified according to the revised epicenters. The first year of the data base is selected as 1816 to exclude three great New Madrid earthquakes, because the average recurrence time of such events is certainly greater than the 187 years (1804-1990) in which the data were collected. The events of $m_b$ less than 1.0 are also excluded. The data is then divided into two parts. All the instrumental data with $1 \leq m_b \leq 4$ is labeled as data set 1 (2201 events). The second part of the data is a combination of historical data and instrumental data of larger magnitude ($m_b \geq 3$), denoted as data set 2 (202 events). By applying the least square method to data set 1, the frequency-magnitude relationship is determined as

$$\log N = 3.19 - 0.91\, m_b$$

Similarly, the frequency-magnitude relationship determined from data set 2 is

$$\log N = 2.47 - 0.78\, m_b$$

To combine the results from these two data sets, the b value of 0.91 obtained from instrumental data (data set 1) is considered as correct. Then, a point ($m_{bs}$, $N_s$) is selected from data set 2 so that the line $\log N = a - 0.91\, m_b$ through this point is the upper bound of data set 2. From a trial and error process, this point is selected as (4.2, 0.21) and the "a" value is determined as 3.15. Therefore, the preferred frequency-magnitude relationship for the NMSZ is

$$\log N = 3.15 - 0.91\, m_b \tag{3}$$

For engineering applications, the recurrence relationship is usually limited by a minimum (lower bound) magnitude $m_{bo}$ and a maximum (upper bound) magnitude $m_{bu}$. For the New Madrid seismic zone, Johnston (1988) determined the minimum and maximum magnitudes as 5.0 and 7.3, respectively. A tremor is denoted as an earthquake if its magnitude is larger than or equal to $m_{bo}$; thus, the occurrence of an earthquake of any size in a source zone $N_o$ is

$$N_o = N(m_{bo}) = e^{\alpha - \beta\, m_{bo}} \tag{4}$$

The probability that an earthquake with magnitude $\leq m_b$ could occur in a source zone is

$$F(m_b) = 1 - \frac{N(m_b)}{N_o} = 1 - \frac{e^{(\alpha-\beta m_b)}}{e^{(\alpha-\beta m_o)}} = 1 - e^{-\beta(m_b - m_{bo})} \quad (5)$$

To satisfy the property that the probability distribution $F(m_{bu})$ should be equal to 1.0, $F^*(m)$ is defined as follows:

$$F^*(m_b) = \frac{F(m_b)}{F(m_{bu})} = \frac{1 - e^{-\beta(m_b - m_{bo})}}{1 - e^{-\beta(m_{bu} - m_{bo})}} \quad (6)$$

then

$$N(m_b) = N_o\,[1 - F^*(m_b)] \quad (7)$$

## ATTENUATION OF GROUND MOTION

The peak ground acceleration is usually attenuated as the epicentral distance increases. In this study, the following attenuation relationship for the NMSZ, as proposed by Nuttli and Herrmann (1984), is used:

$$\log(A_H) = 0.57 + 0.50\ m_b - 0.83\ \log(R^2 + h^2)^{1/2} - 0.00069\ R \quad (8)$$

where $A_H$ is the horizontal PGA (in cm/sec$^2$) averaged from two horizontal components recorded on unconsolidated soil sites; R is the epicentral distance (in km); and h is the focal depth (in km). On the basis of the instrumental data recorded in the NMSZ, the focal depth is taken as 10 km in this study.

## ANNUAL PROBABILITY OF EXCEEDANCE

The occurrence of an earthquake in a seismic zone is usually assumed to follow a Poisson process. The Poisson model is widely used and is a reasonable assumption in regions where data are insufficient to provide more than an estimate of an average recurrence rate (Cornell 1968). According to the Poisson model, the probability that $A_H$ exceeds a specified value a* is

$$P(A_H > a^*) = 1 - \exp[-\nu_A(A_H > a^*)\ t] \quad (9)$$

where t is the time period of interest (one year in this study); $\nu_A(A_H > a^*)$ is the annual occurrence of an earthquake that has $A_H$ exceeding a* and it can be calculated as follows:

$$\nu_A(A_H > a^*) = N_o \sum_i \sum_j P(A_H > a^* \mid m_{bi}, R_j)\, P(R_j)\, P(m_{bi}) \tag{10}$$

where $P(A_H > a^* \mid m_{bi}, R_j)$ is the probability of the horizontal peak ground acceleration exceeding a specified level $a^*$ given an earthquake of magnitude $m_{bi}$ at an epicentral distance of $R_j$; $P(m_{bi})$ is the probability of an earthquake of magnitude $m_b$ occurring in a source zone; and $P(R_j)$ is the probability of an earthquake occurring at the distance $R_j$ from the site. If the range of magnitude between $m_{bo}$ and $m_{bu}$ is divided into small increment $\Delta m_b$, the probability that $m_{bi} \leq m_b \leq m_{bi} + \Delta m_b$ in a source zone can be determined as

$$P(m_{bi}) = F^*(m_{bi}+\Delta m_b) - F^*(m_{bi})$$

$$= \frac{\beta e^{-\beta(m_{bi}-m_{bo})}}{1-e^{-\beta(m_{bu}-m_{bo})}}\, \Delta m_b \tag{11}$$

If the epicentral distance is divided into small increment $\Delta R_j$, the probability $P(R_j)$ can be computed as

$$P(R_j) = \frac{AS_j}{AS} \tag{12}$$

where $AS$ is the total area of a source zone and $AS_j$ is the area of the source zone between $R_j - \frac{\Delta R_j}{2}$ and $R_j + \frac{\Delta R_j}{2}$.

If the $A_H$ value in equation (8) is considered as deterministic, then the probability $P(A_H > a^* \mid m_{bi}, R_j)$ is either one or zero depending on the computed $A_H$ value. If uncertainty in the amplitude of ground motion is included in the analysis, the $A_H$ value is usually taken to be lognormally distributed.

$$P(A_H > a^* \mid m_{bi}, R_j) = 1 - \Phi \left\{ \frac{\ln a^* - [\ln \tilde{A}_H \mid m_{bi}, R_j]}{[\beta_{AH} \mid m_{bi}, R_j]} \right\} \tag{13}$$

where $\Phi[\cdot]$ is the cumulative distribution function of a standard normal variable; $[\tilde{A}_H \mid m_{bi}, R_j]$ is the median value of $A_H$ caused by an earthquake $m_{bi}$ at the epicentral distance $R_j$; and $[\beta_{AH} \mid m_{bi}, R_j]$ is the logarithmic standard deviation of $A_H$.

## SEISMIC HAZARD CURVES

The basic case is defined by assuming no uncertainty in the amplitude of the ground motion. For the basic case, the seismic hazard at the six sites is evaluated by using the procedure described above and the resulting seismic hazard curves are shown in Figure 2.

It is well known that determining seismic source zones, recurrence of earthquakes, amplitude of ground motion, etc., cannot be done precisely without uncertainty. However, a complete uncertainty analysis is beyond the intent of this study. In the following, the effects of uncertainty in the amplitude of ground motion on the resulting seismic hazard curve for site 1 is investigated.

For the case that the variation in $A_H$ is included in the analysis, the $A_H$ value is assumed to be lognormally distributed. The median value of $A_H$ is taken as the value determined from equation (8) and the logarithmic standard deviation of $A_H$ is determined as 0.5. Figure 3 shows the comparison of the seismic hazard curve including uncertainty and the basic-case curve. At the same level of annual probability of exceedance, the PGA value predicted by including uncertainty is larger than the PGA value estimated with no certainty. The difference becomes larger as the annual probability of exceedance becomes smaller.

## PROBABILITY-BASED SCENARIO EARTHQUAKES

Probabilistic seismic hazard analysis is able to determine the annual probability of exceedance of a seismic parameter by considering various sources of uncertainty. However, physical characteristics of the earthquake corresponding to a specified level of probability of exceedance disappear, because the peak ground acceleration is determined from the contribution of various seismic sources. For engineering applications, it is important to establish these physical characteristics such as the earthquake acceleration time history or the response spectra. In this study, the method proposed by Ishikawa and Kameda (1991) is used to determine the hazard-consistent magnitude and hazard-consistent distance for a specified annual probability of exceedance. These quantities are then used to establish the ground motion of the probability-based scenario earthquake that

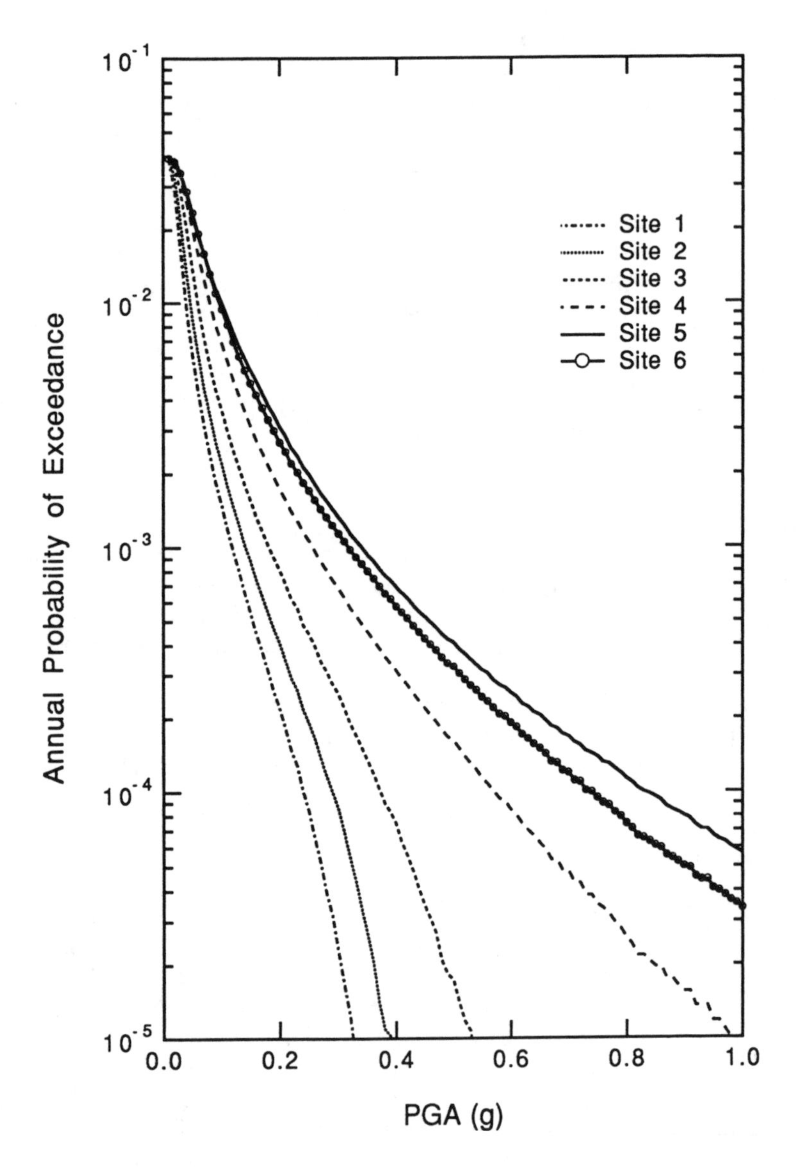


Fig. 2 Seismic Hazard Curves for Six Sites

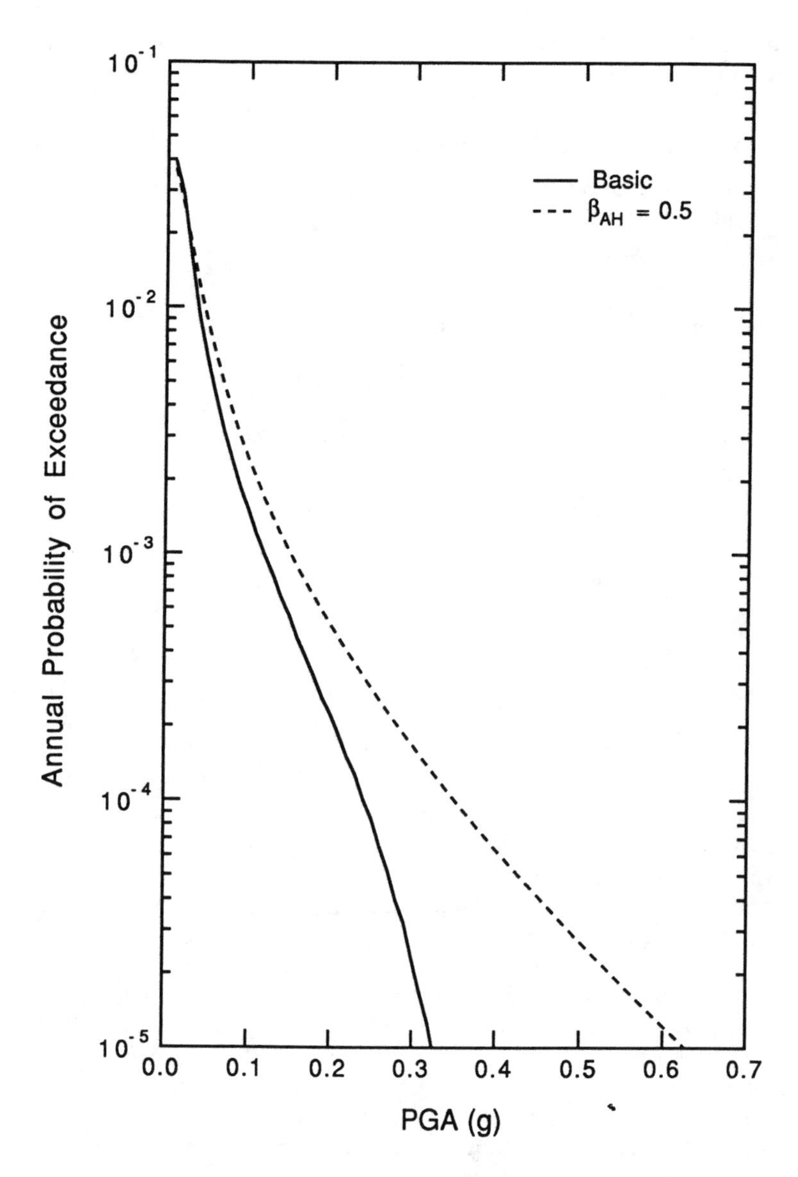


Fig. 3 Effect of Uncertainty in Ground-Motion Attenuation on Seismic Hazard Curve

is the earthquake corresponding to a specified annual probability of exceedance.

The hazard-consistent magnitude $\bar{m}_b$ and hazard-consistent distance $\bar{R}$ are defined as the conditional mean values given the $A_H$ value exceeding the level corresponding to a specified probability of exceedance $a^*(p_o)$.

$$\bar{m}_b(p_o) = E[m_b \mid A_H > a^*(p_o)]$$

$$= \frac{\sum_i \sum_j m_{bi} P(A_H > a^*(p_o) \mid m_{bi}, R_j) P(R_j) P(m_{bi})}{\sum_i \sum_j P(A_H > a^*(p_o) \mid m_{bi}, R_j) P(R_j) P(m_{bi})} \tag{14}$$

and

$$\bar{R}(p_o) = E[R \mid A_H > a^*(p_o)]$$

$$= \frac{\sum_i \sum_j R_j P(A_H > a^*(p_o) \mid m_{bi}, R_j) P(R_j) P(m_{bi})}{\sum_i \sum_j P(A_H > a^*(p_o) \mid m_{bi}, R_j) P(R_j) P(m_{bi})} \tag{15}$$

For site 1, the hazard-consistent magnitudes and hazard-consistent distances corresponding to various probabilities of exceedance are determined and shown in Figure 4. The hazard-consistent magnitude tends to increase as the probability of exceedance decreases, whereas the hazard-consistent distance decreases as the probability of exceedance decreases.

For all six sites, Table 2 shows the hazard-consistent magnitudes and hazard-consistent distances corresponding to the probability of exceedance of 1/1000 per year. The hazard-consistent magnitude for sites 1, 2 and 3 has the same value 6.8; thus, the hazard-consistent distances for these three sites are used to establish a scenario earthquake of magnitude 6.8 at source I (Figure 5) for evaluating seismic hazards at sites 1, 2 and 3. Similarly, we can define a scenario earthquake of magnitude 6.6 at source II for evaluating the seismic hazard at sites 4, 5, and 6 (Figure 5).

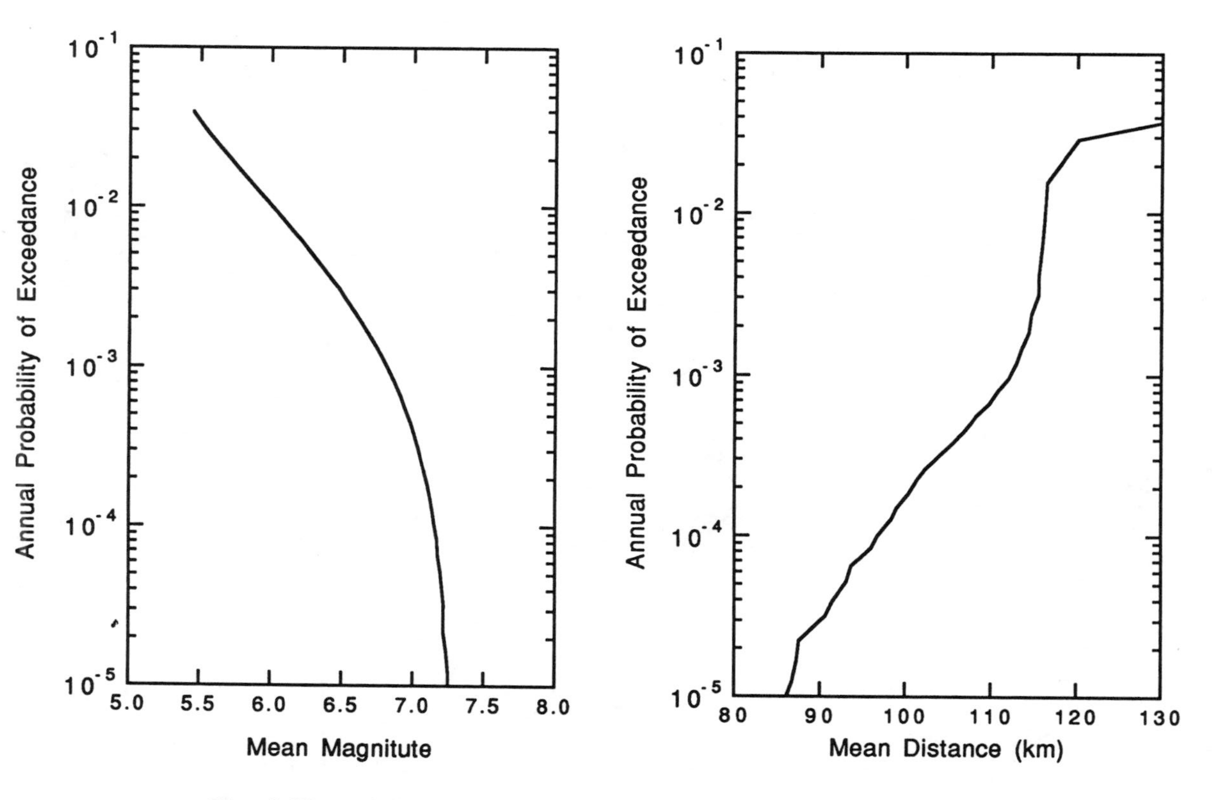


Fig. 4 Hazard-Consistent Magnitudes and Distances (Site 1)

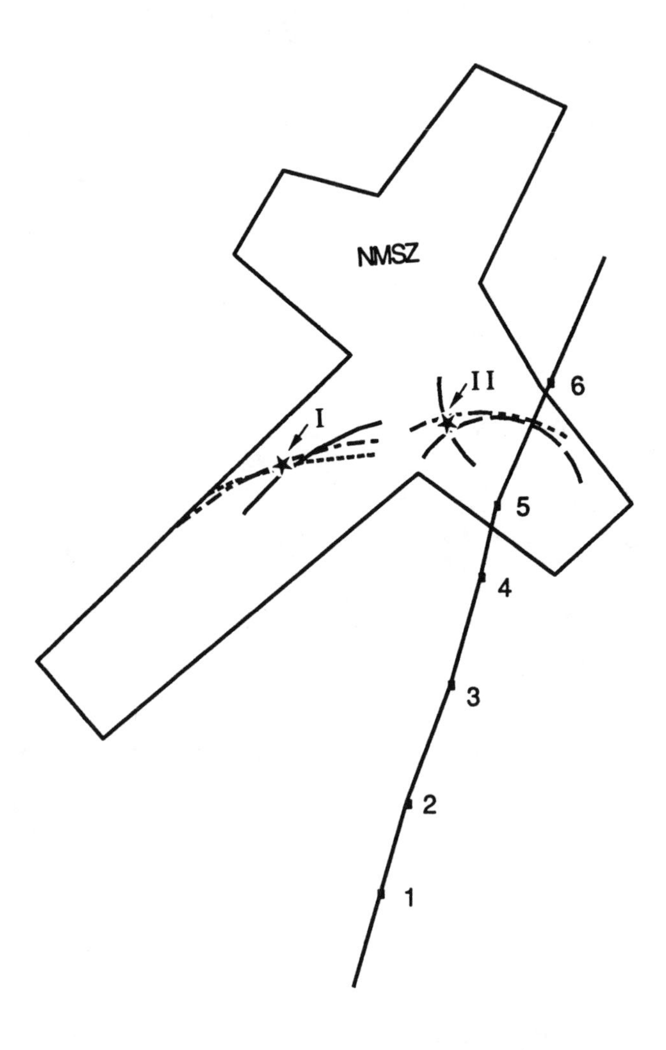


Fig. 5 Probability-based Scenario Earthquakes

Table 2 Mean Magnitudes and Distances (Probability of Exceedance of 1/1000 per year)

| Site | $\bar{m}_b$ | $\bar{R}$ (km) |
|---|---|---|
| 1 | 6.8 | 112 |
| 2 | 6.8 | 93 |
| 3 | 6.8 | 70 |
| 4 | 6.7 | 42 |
| 5 | 6.5 | 23 |
| 6 | 6.6 | 29 |

## CONCLUSIONS

In this study, an integrated approach is used to evaluate the seismic hazards at six sites where pipeline 22 crosses the major rivers in West Tennessee. First, a seismic hazard curve is established by considering uncertainties in seismic zones, maximum magnitude, and amplitude of ground motion. Then, the hazard-consistent magnitude and hazard-consistent distance are determined for a specified annual probability of exceedance. By using these quantities, the ground motion of the probability-based scenario earthquake, for example, the synthetic acceleration time histories can be established. The use of probability-based scenario earthquake for seismic vulnerability assessment has the following advantages: (1) the scenario earthquake has a well-defined annual probability of exceedance; it is unnecessary to specify it arbitrarily or select it as the most conservative value; (2) the characteristics of the scenario earthquake can be realistically established; and (3) the spatial effects of seismic hazards on the lifeline facilities can be assessed.

The seismic hazard curves and probability-based scenario earthquakes presented in this paper are established based on the best-estimate values of the seismic parameters. For the case that the seismic hazards are required to be estimated rigorously, modeling uncertainties in all the steps of the probabilistic seismic hazard analysis need to be considered.

## ACKNOWLEDGMENTS

This paper is based on research sponsored by the National Center for Earthquake Engineering Research (NCEER) under contract number NCEER-90-3012 (NSF grant number ECE-86-07591) and by the U.S. Geological Survey (USGS) under award number 14-08-0001-G2000. Any opinions, findings, and conclusions expressed in the paper are those of the author and do not necessarily reflect the views of NCEER, NSF, or USGS of the United States.

## REFERENCES

Beavers, J.E., Domer, R.G., Hunt, R.J., and Rotty, R.M. (1986), "Vulnerability of Energy Distribution systems of an earthquake in the Eastern United States - An Overview," American Association of Engineering Societies, Washington, DC.

Cornell, C.A. (1968), "Engineering Seismic Risk Analysis," *Bulletin of the Seismological Society of America,* 58, 1583-1606.

Gordon, D.W. (1988), "Revised Instrumental Hypocenters and Correlation of Earthquake Locations and Tectonics in the Central United States," Professional Paper 1364, U.S. Geological Survey.

Gutenberg, B., and Richter, C.F. (1944), "Frequency of Earthquakes in California," *Bulletin of the Seismological Society of America,* 34(4), 185-188.

Ishikawa, Y. and Kameda, H. (1991), "Probability-Based Determination of Specific Scenario Earthquakes," Proceedings of the Fourth International Conference on Seismic Zonation, Stanford, CA, II, 3-10.

Johnston, A.C. (1988), "Seismic Ground Motions in Shelby County, Tennessee, Resulting from Large New Madrid Earthquakes," CERI Technical Report, Center for Earthquake Research and Information, Memphis State University, Memphis, TN.

Johnston, A.C., and Nava, S.J. (1990), "Seismic-Hazard Assessment in the Central United States," in *Neotectonics in Earthquake Evaluation,* Krinitzsky, E.L., and Slemmons, D.B., eds., Geological Society of America, Reviews in Engineering Geology, VIII, 47-57.

Nuttli, O.W. and Herrmann, R.B. (1984), "Ground Motion of Mississippi Valley Earthquakes," *Journal of Technical Topics in Civil Engineering,* 110(1), 54-69.

# Seismic Hazard Analysis for Crude Oil Pipelines in the New Madrid Seismic Zone

Michael J. O'Rourke[1]

Abstract

The seismic vulnerability of three crude oil pipelines in and around the New Madrid Seismic Zone (NMSZ) is evaluated for a repetition of the 1811–12 New Madrid events. The seismic hazard considered is distributed permanent ground deformation (PGD) due to liquefaction and subsequent lateral spreading. Wave propagation effects, and abrupt PGD at fault crossing or at the margins of a lateral spread zone are not considered.

## Introduction

In 1990 the National Center for Earthquake Engineering Research (NCEER) published a pilot study on seismic vulnerability of crude oil pipeline in the NMSZ. In the Ariman et al. (1990) report, the seismic hazard due to liquefaction was quantified, component reliabilities were evaluated and the overall vulnerability of three crude oil lines was determined. Herein the NCEER study is updated by incorporating new information from Japan on the spatial extent of PGD zones and more detailed component reliability analyses.

As in the NCEER study, wave propagation effects are neglected. Most welded steel pipelines have performed well when subject only to the wave propagation hazard. In point of fact, the author is aware of only one case history of wave propagation damage to a relatively corrosion free welded steel. The damage in that unusual case has been attributed by O'Rourke and Ayala (1990) to very soft soils, large amplification and the presence of surface waves at the site in Mexico City.

In addition, abrupt PGD at a fault crossing or at the margins of a lateral spread is neglected. Although this type of ground movement could result in pipeline damage, the behavior is strongly influenced by site specific information, specifically the pipeline anchor length at each side of the fault. Such detailed

---

[1]Professor, Civil & Environmental Engineering, Rensselaer Polytechnic Inst., Troy, NY 12181.

information is not available and hence abrupt PGD is considered beyond the scope of this paper.

## Crude Oil Transmission System

Three pipelines which transmit crude from Texas, Louisiana and Oklahoma to refineries in the Midwest are of particular interest because they pass near the NMSZ. These lines are Pipeline 22, a 40 in. diameter line operated by Shell Pipeline Corporation, Pipeline 66, a 22 in. line operated by the Mid–Valley Pipeline Company, and Pipeline 68, a 20 in. line operated by the Mobile Pipeline Company. The annual capacities of these lines are 230, 98 and 54 million barrels respectively.

The 40 in. Shell system (Pipeline 22) contains 633 miles of crude oil pipeline. The mainline pipe is API 5LX–X52 with a minimum yield strength of 52 ksi. The 20–in. Mobile system contains 648 miles of mainline and 9 miles of 20–in. loop crossing the Arkansas and Mississippi Rivers. The maximum discharge pressures at the booster stations range between 800 and 950 psi. Based on this pressures for the 20 inch line, we assume on average wall thickness of 0.5 in for the 20 and 22 in diameter lines and 0.75 in. for the 40 in. line.

## New Madrid Seismic Zone Hazard

Seismic hazards for the three pipelines of interest were determined by Ariman et al. (1990) for an assumed repetition of the 1811–12 New Madrid events. That evaluation is used herein to characterize liquefaction and associated ground deformation hazards. The 1811–12 New Madrid sequence consisted of four main events. Three of these earthquakes had estimated surface wave magnitudes Ms = 8.4 to 8.7, while the event of 12/16/1811 was of a slightly smaller size.

The liquefaction hazard evaluation was based on the work by Algermissen and Hopper (1985) and Obermeier and Wingard (1985). Algermissen and Hopper replaced the 1811–12 earthquake sequence by the similar but somewhat more conservative assumption of an $M_s = 8.6$, maximum MMI = XI earthquake occurring anywhere along the NMSZ.

Obermeier and Wingard (1985) took these MMI zones defined by Algermissen and Hopper and superimposed on all zones of MMI $\geq$ IX the areas containing liquefiable soils. For example, in zones of MMI IX, clean sand deposits were considered liquefiable if they had a median SPT $\leq$ 12 blows/ft. at a depth of 12 to 20 ft. and the water table was shallow. The three pipelines were carefully superimposed on 1:1,000,000 scale maps prepared for the states of interest by Obermeier and Wingard. The amount of expected permanent ground deformation (PGD) was based on the work by Youd and Perkins (1987) and Turner and Youd (1987), in which they developed the Liquefaction Severity Index (LSI). LSI is defined as the amount of ground failure displacement, in inches, associated with lateral spreads on gently sloping ground (slope between 0.5 and 5%) and the worst possible soil conditions. LSI is defined between 0 and 100, and ground displacements larger than 100 inches are assigned the maximum LSI= 100. Using data available for the 1811–12 New Madrid earthquakes, Turner and Youd proposed the following relation between LSI and distance R where R is defined as the shortest horizontal distance in kilometers to the closest point of the line source. The equation for New Madrid 1811–12 events is

$$\log(\mathrm{LSI}) = 4.252 - 1.276 \log(\mathrm{R}) \tag{1}$$

Contours of equal LSI for the 1811–12 event, calculated using Equation 1, were plotted on top of Obermeier and Wingard maps. These LSI contours are shown for Western Tennessee in Figure 1 along with pipelines 22 and 66.

Table 1 presents a summary of the seismic hazard information for all three pipelines from the NCEER report. For each pipeline, the miles of pipe exposed to various LSI intensities are listed.

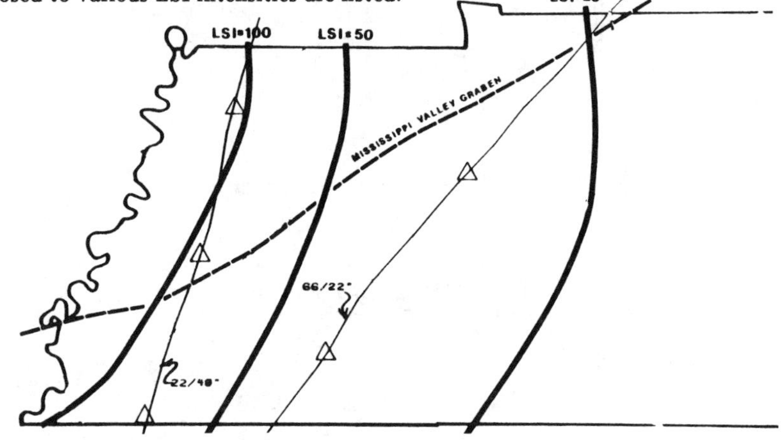


Figure 1 LSI Contours for Western Tennessee for the 1811–12 Earthquakes Using Equation 1.

| | LSI = 10 to 20 | LSI = 20 to 50 | LSI = 50 to 100 | TOTAL |
|---|---|---|---|---|
| PIPELINE 22 | 6 | 29 | 31 | 66 |
| PIPELINE 66 | 30 | 47 | 0 | 77 |
| PIPELINE 68 | 3 | 42 | 20 | 65 |

Table 1 Summary of Pipeline Exposure to Liquefiable Soil and LSI = 10 to 100 Inches for a Recurrence of the 1811–12 New Madrid Earthquake Sequence.

Spatial Extent of PGD

Buried Pipeline response to distributed PGD is a function of the PGD magnitude, the spatial extent of the PGD zone and the pattern of ground movement along the pipeline route. In the 1990 NCEER report, the expected spatial extent of PGD zones was based upon an informal survey of existing expert opinion. Herein more recent data from Japan is used.

Using data from the 1964 Niigata Earthquake and the 1983 Nihonkai Chubu Earthquake, Suzuki and Masuda (1991) identify a number of PGD patterns. Figure 2 shows a typical pattern of longitudinal PGD which induces axial compression in a buried pipeline while Figure 3 shows a typical pattern of transverse PGD which induces bending in a buried pipeline. Figure 4 is a plot of observed values for the magnitude $\delta$ and length L for longitudinal PGD.

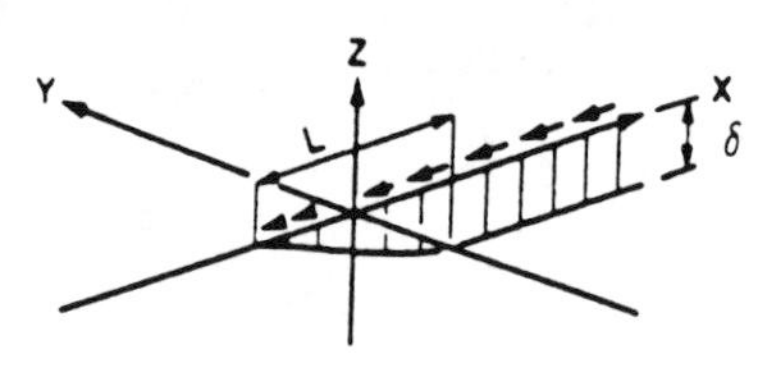


Figure 2 Ramp pattern of longitudinal PGD.

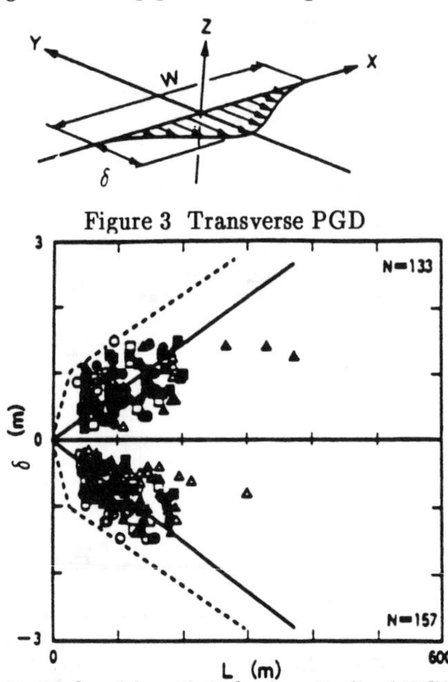


Figure 3 Transverse PGD

Figure 4 Magnitude $\delta$ and length L for longitudinal PGD (after Suzuki and Masuda 1991).

Note in Figure 4 that the average ground strain, $\alpha = \delta/L$, for both tensile (data points above the $\delta = 0$ line in Figure 4) and compressive deformations generally falls in the range $0.002 \lesssim \alpha \lesssim 0.03$ with typical values being 0.007 or 0.008.

Similar plots of the magnitude $\delta$ and width W for transverse PGD suggest the ratio $\delta/W$ generally falls in the range $1/1000 \lesssim \delta/W \lesssim 1/100$ with an average value of $\delta/W \simeq 1/350$.

## Response to Transverse PGD

As noted previously, pipeline response to PGD is a function of the orientation of the pipeline with respect to the direction of ground motion. In general, a pipeline would be exposed to some combination of transverse PGD and longitudinal PGD. For transverse PGD the soil movement is perpendicular to the pipeline axis, while for longitudinal PGD the soil movement is parallel to the pipeline axis. For transverse PGD we neglect any abrupt displacement at the margins of the lateral spread zone and model the lateral ground displacement in Figure 3 by a sinusoidal

$$y(x) = \frac{\delta}{2}\left[1 - \cos\frac{2\pi x}{W}\right] \tag{2}$$

Assuming conservatively that the lateral displacement of a buried pipeline exactly matches the ground displacement pattern given in Equation (2), the bending strain in the pipeline becomes

$$\epsilon_b = \pi^2 \frac{\delta\,\phi}{W^2} \tag{3}$$

where $\phi$ is the pipe diameter. O'Rourke and Nordberg (1991) have shown that equation (3) matches well the results of finite element analysis by others.

## Response to Longitudinal PGD

For a continuous pipeline free of elbows or bends subject to longitudinal PGD, movement and stress in the buried pipeline is due to forces at the soil–pipeline interface. It is assumed that the pipeline is located in a stiff non–liquefied layer which moves due to liquefaction of an underlying layer. This assumption is based on Fuller (1912), who states that in most cases, the depth of fissures is limited to the hard clayey zone extending 10 to 20 feet down from the ground surface to the quicksand. Herein, we conservatively model the interface as a slider with a friction force per unit length at the soil–pipeline interface $f_m$ taken as coefficient of friction times the product of the pipe circumference and the average of the vertical and the horizontal pressure on the pipeline, that is:

$$f_m = \mu \cdot \gamma H \cdot \frac{(1+K_0)}{2}\,\pi\phi \tag{4}$$

where $\mu$ is the coefficient of friction at or near the soil/pipeline interface, $\gamma$ is the unit weight of the soil, H is the depth to the pipe center line, and $K_0$ is the coefficient of lateral earth pressure.

Elhmadi and O'Rourke (1989) suggest $\mu = 0.9 \tan \phi_s$ for most pipe where $\phi_s$ is the angle of shearing resistance of the soil. Because of the backfilling and compaction of the soil around the pipeline, $K_0 = 1.0$ is recommended as a conservative estimate under most conditions of pipeline burial.

For a buried pipeline without bends or elbows subject to a ramp pattern of longitudinal PGD, two configurations are possible as shown in Figure 5(a) and 5(b). For both configurations, it is assumed that slip occurs between the soil and pipeline over a length $L_e$ on either side of the center of the lateral spread zone. Beyond $L_e$ the relative displacement between the soil and pipe is assumed to be zero.

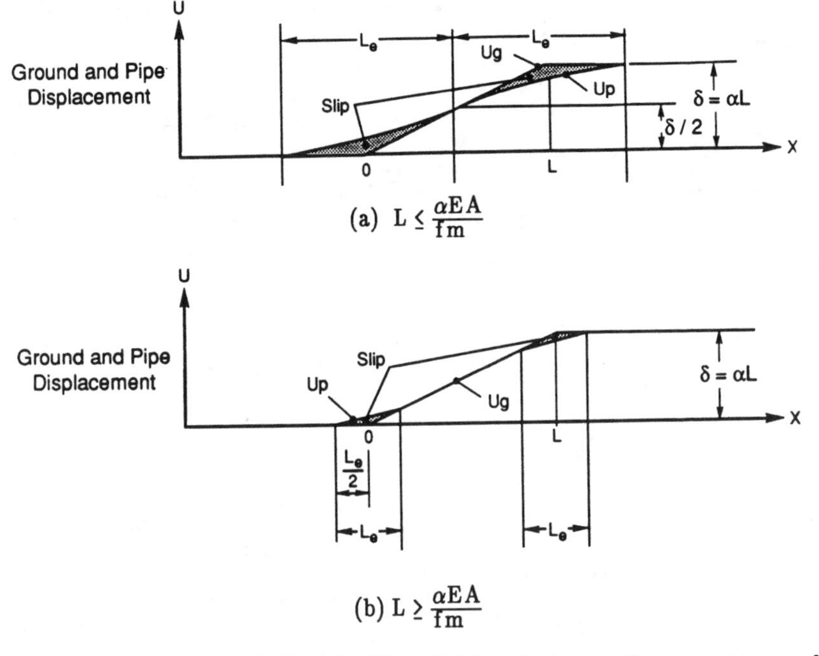


Figure 5 Model of Buried Pipe Subjected to a Ramp pattern of Longitudinal PGD ($U_g$ = ground movement, $U_p$ = axial displacement of pipe).

In Figure 5(a) the slip region extends all the way to the center of the lateral spread zoned. In Figure 5(b), the slip region exists over a limited length near the ends of the zone. For the configuration in Figure 5(a), the maximum pipe strain ($\epsilon < \alpha$) can be determined by noting that the pipe displacement at the center of the lateral spread zone is equal to the displacement of the soil. This displacement can be calculated by integration of the pipe strain over the slip length $L_e$.

$$\alpha L/2 = \int_0^{L_e} \frac{f_m S}{EA} ds = \frac{f_m L_e^2}{2EA} \tag{5}$$

where E is the pipe modulus of elasticity and A is the pipe cross–sectional area. Equation (5) can be solved for $L_e$. The maximum force in the pipe is equal to $f_m$ time $L_e$ since it is assumed that the pipe is not stressed beyond $L_e$. The maximum pipe strain, is then $f_m L_e/EA$ or

$$\epsilon = \sqrt{\frac{\alpha L \ f_m}{EA}} \leq \alpha \tag{6}$$

The relationship in Equation (6) can be simplified by introducing an embedment length $L_{em}$ defined as the distance over which the constant slippage force $f_m$ must act to induce a pipe strain equal to the ground strain

$$L_{em} = \alpha EA/f_m \tag{7}$$

and

$$\epsilon = \begin{cases} \alpha \sqrt{\dfrac{L}{L_{em}}} & L < L_{em} \\ \alpha & L > L_{em} \end{cases} \tag{8}$$

Another possible pattern of longitudinal PGD is Block PGD. This corresponds to a mass of soil having length L, moving as a rigid body down a slight incline. The soil displacement and soil strain at either side of the lateral spread zone are zero, while the displacement of the soil within lateral spread zone is a constant value $\delta$. A ground crack occurs at the head of the slide and a compression mound at the toe.

A Block PGD pattern might be an appropriate model for the ground movement observed by Hamada et al. (1986) in Noshiro City after the 1983 Nihonkai–Chubu earthquake and shown in Figure 6. In Figure 6 the PGD was horizontally to the right, and the height of the open circle indicates the magnitude of the horizontal movement.

For a buried pipeline without bends or elbow, subject to a block pattern of longitudinal PGD, two configurations are possible as shown in Figure 7. For both configurations the pipe strain is largest at the head and toe of the PGD zone. For small length, L, of the PGD zone shown in Figure 8(a), there is slippage at the soil pipeline interface over the whole length of the PGD zone but the maximum pipe displacement at the center of the zone is less than the ground movement $\delta$. Since the force in the pipe by symmetry is zero at the center of the zone, a slip zone of length L/2 exists before the head of the zone and beyond the toe of the zone.

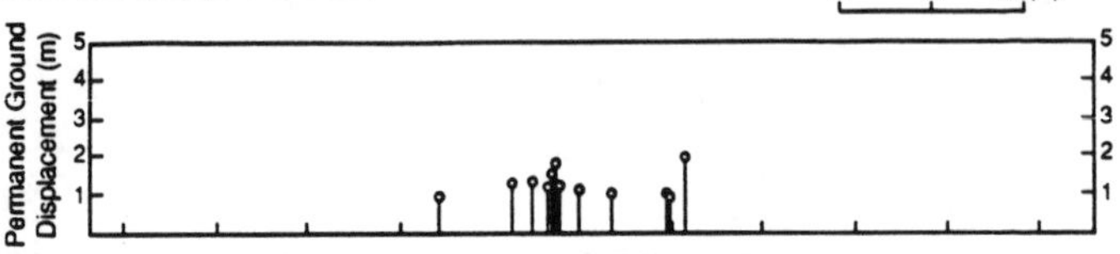


Figure 6 PGD patterns observed in Noshiro City after the 1983 Nihonkai–Chubu Earthquake (after Hamada et al., 1986).

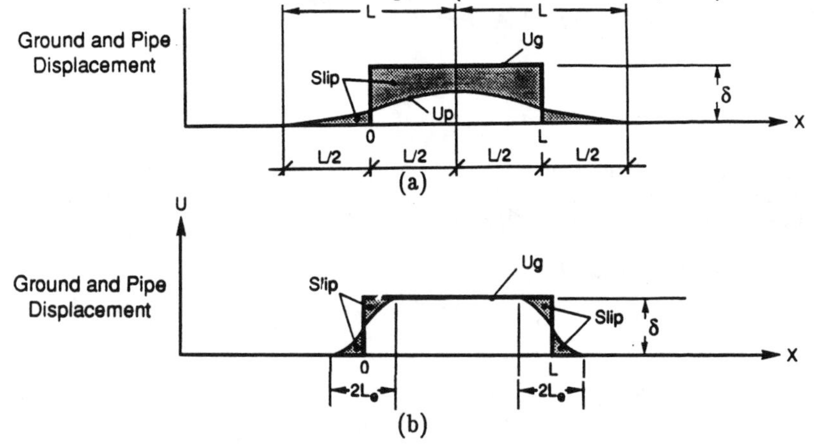


Figure 7 Model of Buried Pipe Subjected to Rigid Block PGD ($U_g$ = ground movement, $U_p$ = axial displacement of pipe).

For large length L of the PGD zone shown in Figure 7(b), a slip zone of length $L_e$ exists at either side of both the head and toe. However the pipe displacement matches ground displacement near the middle of the PGD zone. Assuming that L is large as shown in Figure 7(b), an equation for $L_e$ can be determined by enforcing continuity of the pipe. The displacement due to the stretching of the pipe over the slip region must equal the displacement of the soil, $\delta$, since relative displacement between the soil and pipe are assumed to be zero beyond the slip region. Noting that the length $L_e$ can not exceed L/2, we have

$$\frac{f_m(L_e^2 + L_e^2)}{2EA} = \delta \tag{9}$$

and

$$L_e = \sqrt{\frac{AE\delta}{f_m}} \leq L/2 \tag{10}$$

The maximum force in the pipe is equal to the force per unit length times $L_e$. Upon substitution of the constitutive equations, the maximum pipe strain, $\epsilon$, due to Rigid Block PGD is:

$$\epsilon = \sqrt{\frac{f_m \ \delta}{EA}} \leq \frac{f_m \ L}{2EA} \tag{11}$$

Defining an "equivalent" ground strain for this pattern as

$$\alpha = \delta/L \tag{12}$$

and utilizing the embedment $L_{em}$ from Equation (7) we get

$$\epsilon = \begin{cases} \dfrac{\alpha L}{2L_{em}} & L < 4\,L_{em} \\ \alpha\sqrt{\dfrac{L}{L_{em}}} & L > 4\,L_{em} \end{cases} \tag{13}$$

Nordberg (1991) also considered other idealized patterns of longitudinal PGD. Figure 8 is a plot of normalized pipe strain as a function of the normalized length of the PGD zone for four idealized longitudinal PGD patterns.

The relationships in Figure 8 are for a pipeline free of bends or elbows which would tend to anchor it to the soil, that is slippage is allowed based upon Equation (4). However at a river crossing as shown in Figure 9, we assume that the pipeline is anchored to the displaced soil mass due to the presence of bends or elbows in a vertical plane.

Using arguments similar to those for Equations (5) and (6)

$$\delta = \int_0^{L_e} \frac{f_m S}{EA}\, dS = \frac{f_m L_e^2}{2EA} \tag{14}$$

and the pipe strain at the edges of the displaced soil mass is

$$\epsilon = \sqrt{\frac{2\delta f_m}{EA}} \tag{15}$$

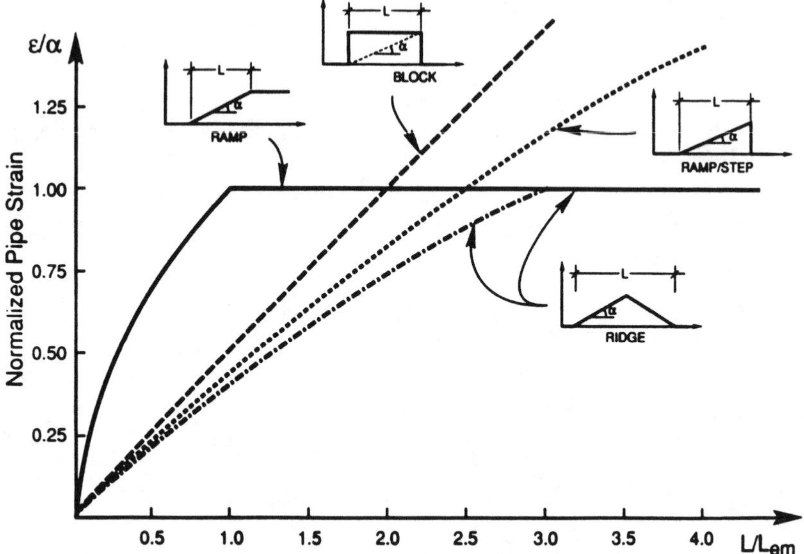


Figure 8 Normalized pipe strain for various patterns of longitudinal PGD (after Nordberg, 1991).

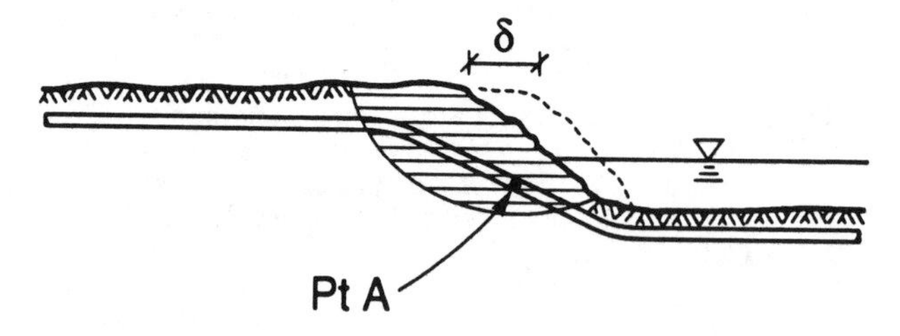


Figure 9 Elevation View of Buried Pipelines at River Crossing Subject to Longitudinal PGD.

compressive strain under the river and tensile to the left of the bank in Figure 9.

Component Reliability, Transverse PGD

For buried pipelines, failure is assumed to occur when the compressional strain in the pipe equals yield. That is, the yield strain is assumed to be equal to the strain for local buckling, which leads to wrinkling and potential tearing of the pipe wall. Hence from Equation (3), failure is assumed to occur for distributed transverse PGD when the width of the lateral spread zone

$$W < \sqrt{\frac{\pi^2 \delta \phi}{\epsilon_y}} \tag{16}$$

where $\epsilon_y$ is the yield strain for the pipe taken herein as 52 ksi/E = 0.0018. For $\delta = 10$ in., the critical widths are 89, 93, and 126 feet for the 20, 22 and 40 in. diameter pipes respectively. For $\delta = 40$ and 100, the critical widths are less than 253 and 399 feet respectively for all pipe diameters. As noted previously, data analyzed by Suzuki and Masuda (1991) suggest that the ratio $\delta/W$ generally fall in the range $1/1000 \leq \delta/W \leq 1/100$ with an average value of $\delta/W \simeq 1/350$. For $\delta$ = LSI = 40 and 100 in., the minimum width of the lateral spread zone W would be 333 ft and 833 ft. Since these expected minimum widths are larger than the critical widths cited above, the pipelines would survive distributed transverse PGD for $\delta$ = LSI = 40 and 100. For $\delta$ = LSI = 10 in., the expected minimum and average widths of a distributed transverse PGD zone are 83.3 ft and 292 ft respectively. Assuming that the average width is the median and that the distribution of widths between the minimum and average (median) value is uniform, the probability of failure for a 22 in. diameter pipe because

$$\text{Prob(fail)} = \text{Prob}(W < 93) = 0.5\,\frac{(93-83.3)}{(292-83.3)} = 0.023$$

The corresponding probabilities of failure for 20 and 40 in. diameter pipes are 0.0135 and 0.102 respectively for distributed transverse PGD with $\delta$ = LSI = 10 in.

At first glance it may seem strange that the probability of failure for distributed transverse PGD is a decreasing function of LSI. Note however that if W is linearly proportional to $\delta$, as suggested by the Japanese data, then Equation (3) indicates that the pipe strain is inversely proportional to $\delta$.

Component Reliability, Longitudinal PGD

Figure 8 indicates for short lengths of the PGD zone ($L < 2L_{em}$), a ramp pattern of longitudinal PGD gives the largest strain in a pipe without bends or elbows. That pattern will conservatively be assumed herein. For H = 5 ft, $\gamma = 105$ pcf and $\mu = 0.60$ corresponding to noncohesive backfill with $\phi_s = 34°$, equation 6 given $\epsilon = 0.001$ for $\delta = \alpha L = 10$ in and $\epsilon > 0.002$ for $\delta = 40$ or 100 in for the 40 in diameter pipeline. Since the assumed failure criterion is a pipe strain of 0.0018, the 40 in "bend free" line survives LSI = 10 but fails for LSI = 40 or 100. It can be shown that this conclusion also applies for the 20 and 22 in lines without bends or elbows.

For a line with bends or elbows such as at a river crossing shown in Figure 9, the equivalent burial depth of the compression portion under the river would be larger than the H = 5 ft assumed above. Taking H > 10 ft. for the river crossing case, equation 15 gives $\epsilon > 0.002$ for all three pipe diameters and $\delta$ = LSI $\geq$ 10 in.

Hence the likelihood of failure due to longitudinal PGD is substantially larger than that for transverse PGD with no abrupt displacement at the margins of the lateral spread zone.

Probability of Lateral Spread Affecting Pipeline

Individual lateral spreads have a finite size. Hence, although the pipeline is in a region with a give LSI, there is some probability less than unity that the pipeline will intersect an individual lateral spread. Note in this regard that Obermeier (1985) stated: "In the 1811–12 earthquakes, this type of landslide was extremely common (probably many thousands) and wide spread in the entire region between the Mississippi River and Crowley's Ridge, from Cairo, Illinois to Memphis, Tennessee." For our purposes, it is assumed that 6,000 is a reasonable upper bound for "many thousands". The distance from Crowley's Ridge to the Mississippi River is 34 miles and the distance from Cairo to Memphis is 124 miles. Hence, although the whole area is susceptible to liquefaction, the probability of a lateral spread in any given square mile is about 1.42 lateral spread/sq. mile for a repetition of the 1811–12 events.

Based upon the Suzuki and Masuda (1991) data i.e. $\delta/W \simeq 1/350$, we take an average width of about 1200 ft. corresponding to $\delta = \text{LSI} = 40$. It is further assumed that a pipeline would be significantly influenced by a lateral spread whose center is within 600 ft of the pipeline (i.e. 1200 ft wide influence width). Hence the mean rate of occurrence of potentially damaging lateral spreads is

$$\lambda = 1.42 \frac{\text{spreads}}{\text{sq. mile}} \times \frac{1200 \text{ ft}}{5280 \text{ ft/mile}} = 0.322$$

or about 32% per mile for PGD away from river crossings.

Based upon work by Jibson (1985), Ariman et al. (1990) estimate the probability of a deep slope failure at a high loess river bank to be about 20%.

System Reliability

Lacking site specific information such as ground slope and the slope of the bottom of the liquefaction layer, it is assumed that at locations away from river crossings there is a 50% chance that a pipeline is orientated perpendicular to the direction of ground movement (transverse PGD) and 50% that it is orientated parallel to the direction of ground movement (longitudinal PGD). Hence for any of the three pipe diameters subject to a PGD zone with $\delta = \text{LSI} = 40$ or 100 in., the probability of failure is $0.5(0.0) + 0.5(1.0) = 0.50$. For the 22 in diameter subject to $\delta = \text{LSI} = 10$, the probability of failure is $0.5(.023) + 0.5(0.0) = 0.012$. The corresponding values for the 20 and 40 in lines are 0.007 and 0.05 respectively for $\delta = \text{LSI} = 10$. Using the pipeline exposure data given in Table 1 and the procedures described in detail in the NCEER report and O'Rourke et al. (1992), the reliability of each pipeline for PGD away from a river crossing is given in the first column of Table 2.

| Pipeline | away from river crossing | at river crossing | combined |
|---|---|---|---|
| 22(40") | 0.24 | 0.33 | 0.08 |
| 66(22") | 0.49 | 0.33 | 0.16 |
| 68(20") | 0.25 | 0.33 | 0.08 |

Table 2 Reliability of Pipelines for Recurrence of 1811–12 events.

For the river crossing hazard, O'Rourke et al. (1992) judged that the potential for deep seated landslides exists at a maximum of five river crossing locations for Pipeline 22. It is assumed that similar exposure to PGD at a river crossing applies to Pipelines 66 and 68. Since, as noted previously, the probability of a deep slope failure at a susceptible location is estimated to be 20% and each of the three pipelines with bends or elbows fails for $\delta = \text{LSI} > 10$, the reliability of each pipeline for the river crossing hazard is estimated as $[1.0 - 0.2(1.0)]^5 = 0.33$. These values are listed in the second column of Table 2. Neglecting the potential for seismic damage to pumping stations, which is discussed in detail by Ariman et al. (1990) and O'Rourke et al. (1992), the overall reliabilities for each line is simply the product of the reliabilities for pipe with no bends (away from river crossing) and pipe with bends (at river crossings). This combined reliability for each line is listed in the third column of Table 2. Note that this analysis suggests that Pipelines 22 and 68 are the most vulnerable, each with roughly a 1 in 10 chance of surviving a repetition of 1811–12 events. Pipeline 66 is somewhat less vulnerable.

## Conclusions

The seismic vulnerability of three crude oil pipelines in and around the New Madrid Seismic Zone (NMSZ) was evaluated for a repetition of the 1811–12 New Madrid events. The seismic hazard considered was distributed permanent ground deformation (PGD) due to liquefaction and subsequent lateral spreading. Wave propagation effects, and abrupt PGD at fault crossing or at the margins of a lateral spread zone are not considered. In addition, seismic damage to pump stations was neglected.

The analysis suggests that the probability of pipe failure due to transverse PGD is quite small. The cumulative probabilities of failure due to longitudinal PGD for a pipeline without bends or elbows (away from river crossing) is roughly equal to that for a pipeline with bends (at a river crossing). For pipelines 22 and 68, the estimated probability of surviving a repetition of the 1811–12 event is roughly 1 in 10. Pipeline 66 is somewhat less vulnerable. These reliability estimates are lower than corresponding values in O'Rourke et al. (1992). This difference is due primarily to the use herein of recent results from Japan on the spatial extent of PGD zones and the conservative assumption herein of a ramp pattern of longitudinal PGD away from river crossings.

The study, described above, also highlights the need for more detailed information on lateral spreads, as outlined below:

- Procedures for estimating the expected number of PGD zones as a function of earthquake size and site–to–source distance is needed for typical soil conditions in the U.S.
- The expected range for the widths W and length L of PGD zones, as function of the amplitude of ground movement $\delta$, is needed for the U.S.
- For longitudinal PGD, we need to quantify the distribution among various patterns, that is, what percentage of PGD zones exhibit a ramp pattern, a block pattern, etc.
- For transverse PGD, what percentage of the total lateral ground movement $\delta$ occurs as an abrupt offset at the margins of the PGD zone.

## Acknowledgement

The work presented herein was supported by the U.S. National Center for Earthquake Engineering Research. This support is greatly acknowledged. However, the findings and conclusions are the authors alone and do not necessarily reflect the view of NCEER.

## References

Algermissen, S.T. and Hopper, M.G., (1985), "Maps of Hypothetical Intensities for the Region," included in *U.S. Geological Survey Open–File Report 85–457*.

Ariman, T., Dobry, R., Grigoriu, M., Kozin, F., O'Rourke, M., O'Rourke, T., and Shinozuka, M., (1990), "Pilot Study on Seismic Vulnerability of Crude Oil Transmission Systems," Technical Report NCEER–90–0008, NCEER, Buffalo, N.Y.

ElHmadi, K., and O'Rourke, M., (1989), "Seismic Wave Propagation Effects on Straight Jointed Buried Pipelines," Report No. NCEER–89–0022, NCEER, Buffalo, NY.

Fuller, M., (1912), "The New Madrid Earthquake," *U.S. Geological Survey Bulletin 494*, Washington, D.C.

Hamada, M., Yasuda, S., Isoyama, R., and Emoto, K., (1986), "Study of Liquefaction Induced Permanent Ground Displacements," Association for the Development of Earthquake Prediction, Japan, November, p. 87.

Jibson, R.W., (1985), "Landslides Caused by the 1811–12 New Madrid Earthquakes," Ph.D. Thesis, Stanford University, March.

Nordberg, C., (1991), "Analysis Procedures for Buried Pipelines Subject to Longitudinal Permanent Ground Deformation," M.S. Thesis, Rensselaer Polytechnic Institute, Troy, NY.

Obermeier, S.F., (1985), "Nature of Liquefaction and Landslides in the New Madrid Earthquake Region," *U.S. Geological Survey Open–File Report 85–457.*

Obermeier, S.F. and Wingard, N., (1985), "Potential for Liquefaction in Areas of Modified Mercalli Intensities IX and Greater," *U.S. Geological Survey Open–File Report 85–457*, pp. 127–141.

O'Rourke, M.J. and Ayala, G., (1990), "Seismic Damage to Pipeline: Case Study," *Journal of Transportation Engineering*, American Society of Civil Engineers, Vol. 116, No. 2, March/April 1990.

O'Rourke, M. and Nordberg, C., (1991), "Analysis Procedures for Buried Pipelines Subject to Longitudinal and Transverse Permanent Ground Deformation," Technical Report NCEER–91–0001, *Proc. 3rd Japan–U.S. Workshop on Earthquake Resistant Design of Lifeline Facilities and Countermeasures for Soil Liquefaction*, San Francisco, CA, 439–453.

O'Rourke, M., Shinozuka, M., Ariman, T., Dobry, R., Grigoriu, M., Kozin, F., and O'Rourke, T., (1992), "Study of Crude Oil Transmission System Seismic Vulnerability," accepted for publication, *Earthquake Spectra.*

Suzuki, N. and Masuda, N., (1991), "Idealization of Permanent Ground Movement and Strain Estimation of Buried Pipes," Technical Report NCEER–91–0001, *Proc. 3rd Japan–U.S. Workshop on Earthquake Resistant Design of Lifeline Facilities and Countermeasures for Soil Liquefaction*, San Francisco, CA, pp. 455–469.

Turner, W.G. and Youd, T.L., (1987), "National Map of Earthquake Hazard," *Final Report to USGS of Grant No. 14–08–001–G1187*, Dept. of Civil Engineering, Brigham Young University.

Youd, T.L. and Perkins, D.M., (1987), "Mapping of Liquefaction Severity Index," *Journal of Geotechnical Engineering*, Vol. 113, No. 11, pp. 1374–1392, November.

# Migration of Spilled Oil from Ruptured Underground Crude Oil Pipelines in the Memphis Area

Otto J. Helweg,[1] Fellow, ASCE

Abstract

This study evaluated the impact of an oil pipeline rupture in the recharge area of a West Tennessee aquifer. Two potential break locations were chosen in the alluvial valley of a main river where the probability of a rupture is greatest. The volume and fate of the hydrocarbons were modeled, using MOFAT and ARMOS (two finite element models) and remediation strategies proposed. The results showed a vast volume of oil (2260 $m^3$) would be released in the event of a rupture , but the migration through the porus media would be slow, covering an area of about 50,000 $m^2$ after 60 days. This would give adequate time for remediation measures; however, the amount of BTEX's that would dissolve in the water could reach 276 Kg after 60 days, requiring a significant amount of ground water be treated.

## Introduction

Three different national concerns coalesced to provide the impetus for this study. First, the earthquake in California reminded residents of the Mid South that they live in the vicinity of the second most earthquake hazardous zone in the U.S. Second, a national interest in aquifer contamination by hydrocarbons, sometimes called nonaqueous phase liquids (NAPL), has resulted from numerous leaking underground storage tanks (UST). Third, ruptures of several petroleum and other lines have alerted people to the spill potential they present.

Fig. 1 shows the crude oil pipeline routes overlaid on the earthquake zones of the central U.S. (Beavers, et.al., 1986). Estimating

---

[1]Chair and Prof., Dept. of Civ. Engrg. Memphis State Univ., Memphis, TN 38152

the earthquake hazard to these pipelines has been covered by Beavers, et.al. (1986). Also, Hwang, et.al. (1990) have dealt with the earthquake hazard to municipal water distribution systems along a similar line. Consequently, risk or hazard analysis, though part of this study, will not be covered in this paper.

The interest in NAPL contamination is illustrated by the increasing number of books and papers published on the subject. Several examples of books are Canter and Knox (1986) and Calabrese and Kostecki (1988). Another indication of interest is the attendance at the "Petroleum Hydrocarbons and Organic Chemicals in Ground Water" conference. The concern for ground water protection is illustrated by the Conservation Foundations report (1987).

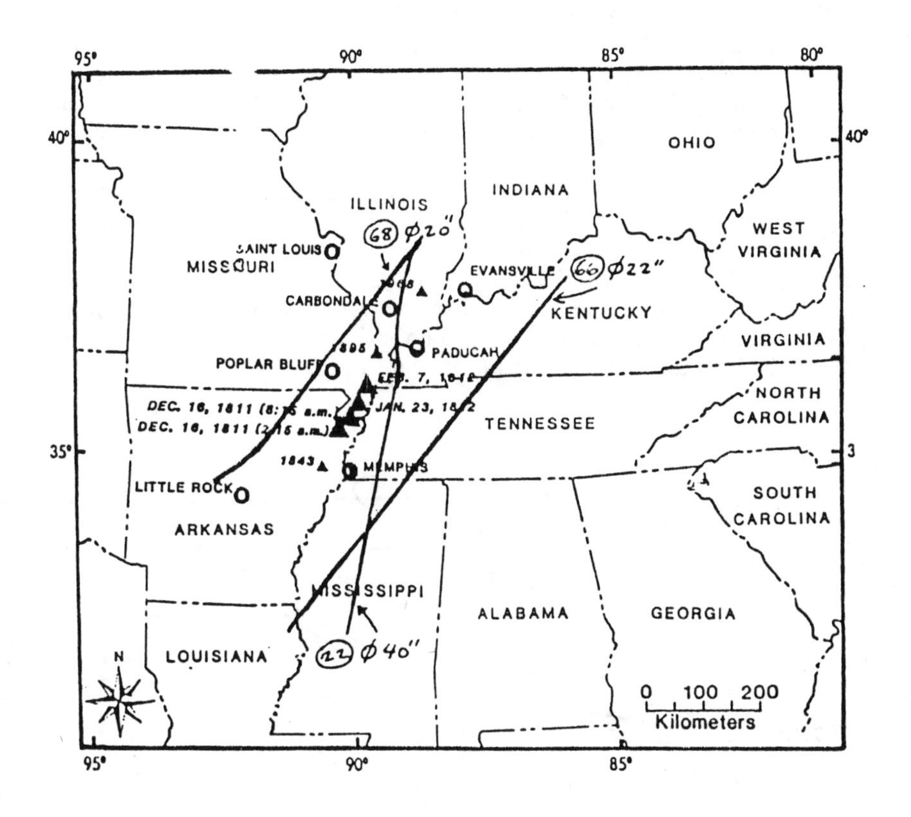


**FIG. 1. Central U.S. Earthquake Zones and Crude Oil Pipelines (Beavers, et.al., 1986)**

Study Site

The proposed rupture zone is where the forty-inch crude oil pipeline crosses the Wolf River at the Shelby/Fayette County line which is south east of Memphis, Tennessee. Here the pipeline not only crosses a stream channel, but almost one mile of wetlands before the land elevation rises out of the flood plain. The pipe is buried about four feet under ground and this area, as many fluvial valleys, is the most susceptible to liquefaction during an earthquake and therefore loss of support to the pipe. Moreover, the pipeline crosses the recharge zone of the Memphis Sands, see Fig. 2

The Memphis Sands yields about 200 mgd (757,400 $m^3$/day). The formation top starts from zero at the recharge area and dips to 500 feet. The aquifer thickness ranges from 500 to 890 feet (Gram and Parks, 1986). MLGW obtains about 98% of their water from this aquifer. The water is excellent for consumption and industrial use. Were this aquifer to become contaminated, it would obviously cause sever problems to the greater Memphis area. Because the area crossed by the pipe line is the recharge zone of the Memphis Sands, any pollution in this region is doubly dangerous.

The pipeline is equipped with booster pumps and cut off valves approximately every 40 miles. In the study area, a pump/valve station is located just inside Mississippi off of Highway 78 with the next station located in Brownsville, Tennessee. The system is remotely controlled at a central station in Louisiana, so the valves and pumps do not have automatic shutoff capabilities. A profile of the pipeline route taken from U.S.G.S. 15 minute quad maps is shown in Fig. 3.

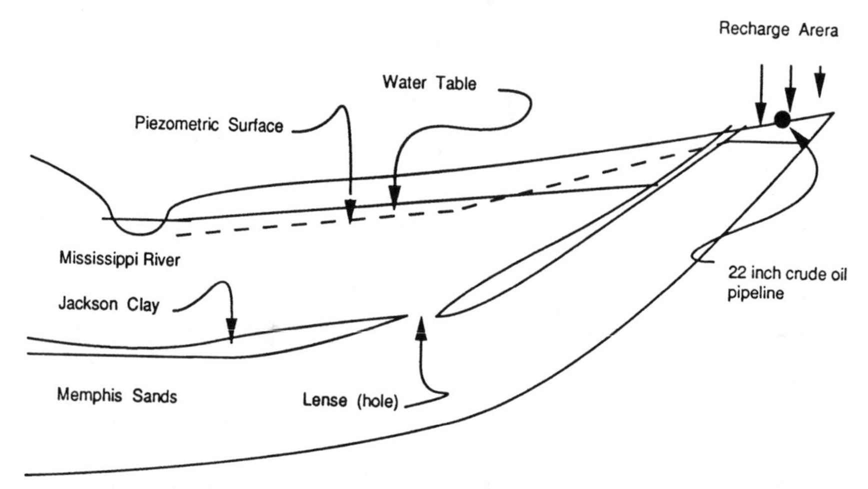


**FIG, 2, Profile of the Memphis Sands and pipe crossing.**

The predominate soil type in the vicinity of the break locations is the Waverly silt loam. This soil is a poorly drained, gray soil usually found on nearly level bottom lands. The Waverly soils consist of recent silty alluvium that is 12 inches to 20 feet thick and generally overlies older, poorly drained, silty material. Permeability is moderately slow with the water table at or near the surface all year. The second major soil group in the area of the potential breaks is the Falaya fine sandy loam. This soil is sandier in the surface layer. Several bore holes were drilled to obtain samples of the alluvial soils. These were analyzed to obtain the hydraulic conductivity, porosity, and other Brooks-Corey parameters as shown in Table 1.

Table 1. Brooks-Corey and other parameters for ARMOS and MOFAT

| Parameters | ARMOS Values | MOFAT Values |
|---|---|---|
| $h_d$ (m) | 0.05 | 0.33 |
| $\lambda$ | 0.55 | 0.79 |
| $\phi$ | 0.35 | 0.35 |
| $\rho ro$ | 0.86 | 0.86 |
| $\eta ro$ | 7.18 | 7.18 |
| $\beta ao$ | 2.40 | 2.40 |
| $\beta ow$ | 1.72 | 1.72 |
| $\varepsilon L$ | N.A. | 0.40 |
| $\varepsilon T$ | N.A. | 0.05 |
| $S_m$ | 0.64 | 0.003 |
| $K_{sat,x}$ (m/d) | 54 | 54 |
| $K_{sat,y}$ (m/d) | 54 | N.A. |
| $K_{sat,z}$ (m/d) | N.A. | 2.3 |

The parameters in Table 1 are as follows: $h_d$ is the Brooks-Corey air entry head, $\lambda$ is the Brooks-Corey constant, $\phi$ is the porosity, $\rho_{ro}$ is the ratio of fluid densities, $\beta ao$ and $\beta ow$ are the air to oil and oil to water scaling factors, $\varepsilon L$ and $\varepsilon T$ are the longitudinal and transverse dispersivities, $S_m$ is the residual saturation, and the K values are for hydraulic conductivity in the x, y, and z directions.

The other data necessary for modeling are the properties of the crude oil. Table 1 also shows some of these parameters. Pipeline 22 normally transmits four types of crude oil, Eugene Island Crude (EIC), Light Louisiana Sweet (LLS), Louisiana Mississippi Sweet (LMS), and Mississippi Canyon Crude (MCC). EIC was chosen for the simulation an a sample of EIC was furnished by the MAPCO Petroleum Corp. and analyzed by Core Laboratories. The API gravity at 15.6 deg. C was 33.6 and the viscosity (SSU) at the same temperature was 58.7.

Finally, in order to simulate the dissolution of various BTEX's, MOFAT required the properties of the various species, see Table 2.

Table 2. Properties of organic components for MOFAT

| Parameters | Benzene | Toluene | O-Xylene | Ethlybenz. |
|---|---|---|---|---|
| $\Gamma_{Bo}$ | 1.5313 | 1.0142 | 3.3250 | 2.6250 |
| $\Gamma_{Ba}$ | 0.2400 | 0.2800 | 0.2200 | 0.3700 |
| $\rho_B$ | 879 | 867 | 861 | 867 |
| $M_B$(kg/mole) | 0.0780 | 0.0920 | 0.1060 | 0.1060 |
| $D_{Bw}$ (m$^2$/day) | 0.00009 | 0.00009 | 0.00009 | 0.00009 |

The parameters in Table 2 are as follows: $\Gamma_{Bo}$ is the equilibrium partition coefficient, $\Gamma_{Ba}$ is the nondimensional Henry's law constant, $\rho_B$ is the density, $M_B$is the molecular weight, and $D_{Bw}$ is the diffusion coefficient. Again, these values were for the Eugene Island Crude (EIC), the crude oil chosen for the simulation.

Procedure

Two breaks in the forty-inch pipeline were modeled, one occurring underneath the Wolf River and the other at the edge of the fluvial valley. The break under the Wolf River entailed surface water contamination; the latter only ground water. The volume of hydrocarbons released after a break was estimated by calculating the volume in the pipeline between the two highest points on either side of the Wolf River. The distance between the two shut-off valves closest to the Wolf Rives is 88,500 meters (55 miles). The volume of oil in this section is 71,700 m$^3$ (1,534,000 ft$^3$). However, as Fig. 3 shows, only the oil between the two highest elevations bracketing the Wolf River would spill from a rupture. This volume was estimated to be 5,600 m$^3$ (200,000 ft$^3$), the volume between the two high points plus the amount pumped through the pipeline during the response time. Of the 5,600 m$^3$ spilled, 2,260 m$^3$ was estimated to infiltrate into the porus media and the rest was assumed to be recovered. This estimate of recovered oil is based on the data collected from the Bemidji spill by the USGS (1984).

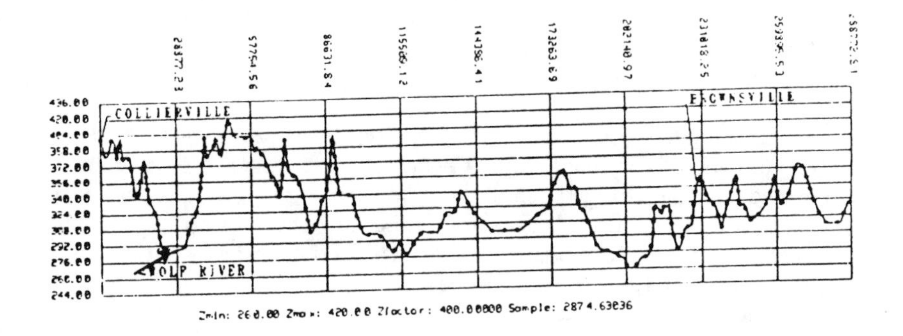

**FIG. 3, Profile of pipeline 22 between Collierville and Brownsville.**

The fate of the released hydrocarbons were modeled by the ARMOS and MOFAT models (Kaluarachchi and Parker, 1989). These finite element models are capable of simulating multiphase flow in two dimensions. They not only defined the probable contaminate flume, but assisted in predicting the effectiveness of the remediation activities. From these results, emergency procedures and preventative measures were suggested to MLGW.

Since ARMOS simulates NAPL in the horizontal plane only, it was used to show the areal extent of the NAPL plume. MOFAT simulates NAPL in the vertical plane and the amount of BTEX's dissolved in the ground water. Since neither model was constructed for an oil pipe break scenario, a number of assumptions and approximations were necessary. For simulating oil infiltration into an initially oil free medium, one cannot use a value of zero for $H_0$, the oil thickness in a monitoring well, because it causes numerical instability, so an practical minimum value of 0.1944 meters was chosen.

The infiltration of oil was simulated by introducing 41 volumetric source-sink nodes in the area of the oil pool caused by the pipe rupture. These may be thought of as infiltration wells. This was a constant rate boundary condition, adjusted by trial and error to achieve a volume of 2260 $m^3$ of oil during the infiltration phase, the estimated amount that would be infiltrated. The results of ARMOS were used to specify the infiltration boundary conditions for MOFAT. The water table boundaries upstream and downstream of the modeled area were constant head of 7.3 and 6 meters respectively to maintain a constant water table hydraulic gradient. A zero flux defined the bottom of the aquifer and vadose zone.

The models did not converge easily; however, since this study, later versions of the models have been released that are more robust. The models have three types of boundary conditions, a constant head (type-1), a constant flux (type-3), and a no-flow boundary (type-2). The models allow for three sub-time periods within the total simulation time, infiltration, distribution, and remediation. The times and types of boundaries for these phases are part of the input data.

## Summary and Conclusions

Figs. 4 and 5 show the plume at two and 60 days respectively resulting from ARMOS. At the end of two days the plume covered an area of about 8,500 $m^2$. After 30 days the plume covered an area of 18,000 $m^2$, but the height of the oil on the water was only 0.24 meters and from that point that was very little spreading because the NAPL gradient was not enough to spread the viscus oil. Figs. 6 and 7 show the plume of Toluene at the end of ten and 120 days resulting from the MOFAT simulation. Toluene was the most soluble of the BTEX's. Fig. 8 shows the solubility of the main BTEX's over time.

A pipeline rupture under the Wolf River presents a totally different problem. Because the NAPL is lighter than water a rupture under the river will produce a floating plume of NAPL on the surface water. The velocity of flow in the Wolf River varies from 0.75 m/s (2.5 fps) to 1.98 m/s (6.5 fps). The travel time of the plume to reach the Mississippi river varies between 7 and 18 hours respectively. The remediation for both these cases would consist of the normal floats, and skimming practices; however, if set up and ready to deploy, the surface spill could be contained within the Wolf River, preventing its spread into the Mississippi.

Timely response to any spill emergency should facilitate an adequate recovery of the pollutants, especially hydrocarbons since they will remain, initially, on top of the water table. It is important to be prepared for disasters that **will** happen. Many people think in terms of "**IF** a disaster happens," rather than "when it happens." This study provided information concerning possible contamination to the Memphis Sands aquifer, extent of surface contamination, and a remediation strategy. Moreover, it determined what, if any, preventative measures should be taken in light of risk. These were mainly, an emergency response plan and tentative remediation plans. A geographical information system (GIS) has been set up utilizing ARC/INFO software to evaluate the research results and facilitate remediation measures.

## References

Beavers, J. E., et. al. (1986). "Vulnerability of Energy Distribution Systems to an Earthquake in the Eastern United States: an Overview." AAES

Calabrese, E. J. and Kostecki, P. T. (1988). **Petroleum Contaminated Soils**. Lewis Publishers, Inc., Chelsia, MI.

Canter, L. W. and Knox, R. C. (1986). **Ground Water Pollution Control**. Lewis Publishers, Inc., Chelsia, MI.

Conservation Foundation (1978). **Groundwater Protection**, The Conservation Foundation, Washington, D.C.

Graham, D. D. and Parks, W. S. (1986). "Potential for Leakage Among Principal Aquifers in the Memphis Area, Tennessee." U. S. Geological Survey, Water Resources Investigations Report 85-4295

Hwang, H., Helweg, O. J., and Smith, J. W. (1990). "Estimating the Earthquake Hazard of Municipal Water Systems." this publication.

Kaluarachchi, J. J. and Parker, J. C. (1989). "An Efficient Finite Element Method for Modeling Multiphase Flow." **Water Resources Research**, Vol. 25, No. 1, Jan.

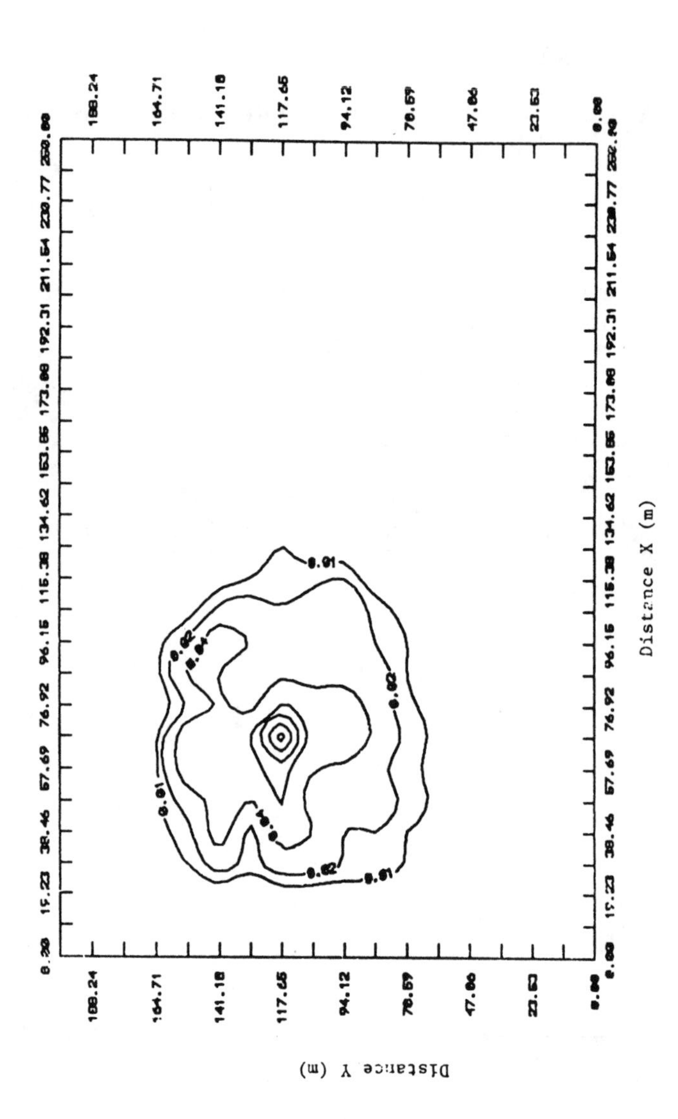


FIG. 4, Migration of EIC oil after 2 days infiltration

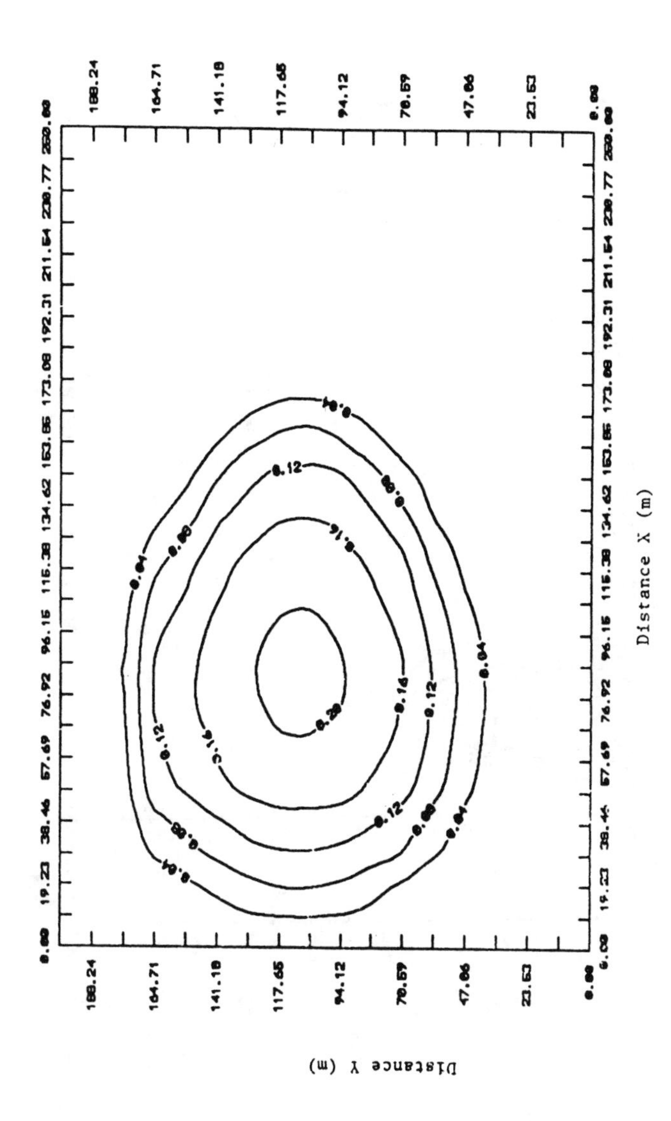


FIG. 5, Migration of EIC oil after 60 days infiltration

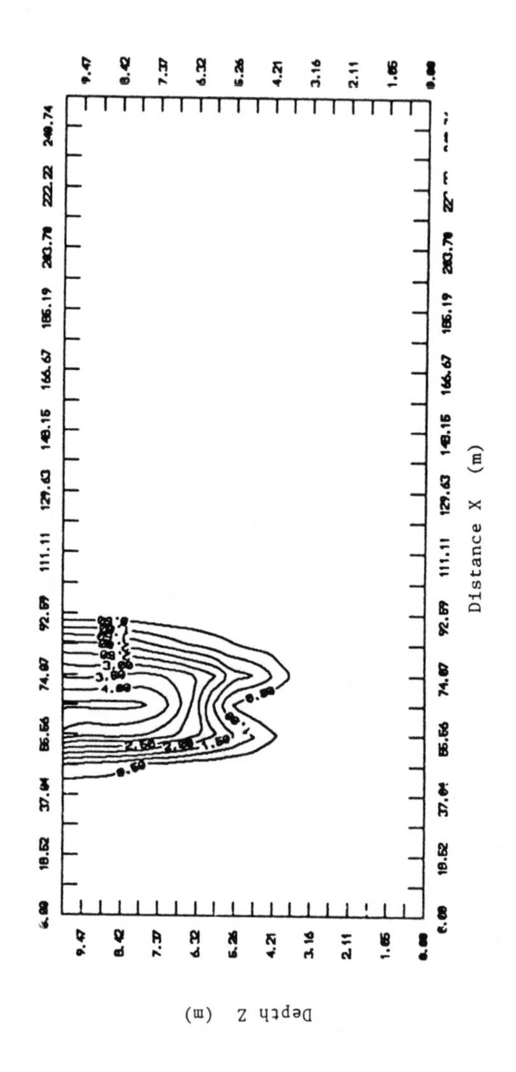


**FIG. 6, Water phase concentration of Toluene plume after 10 days.**

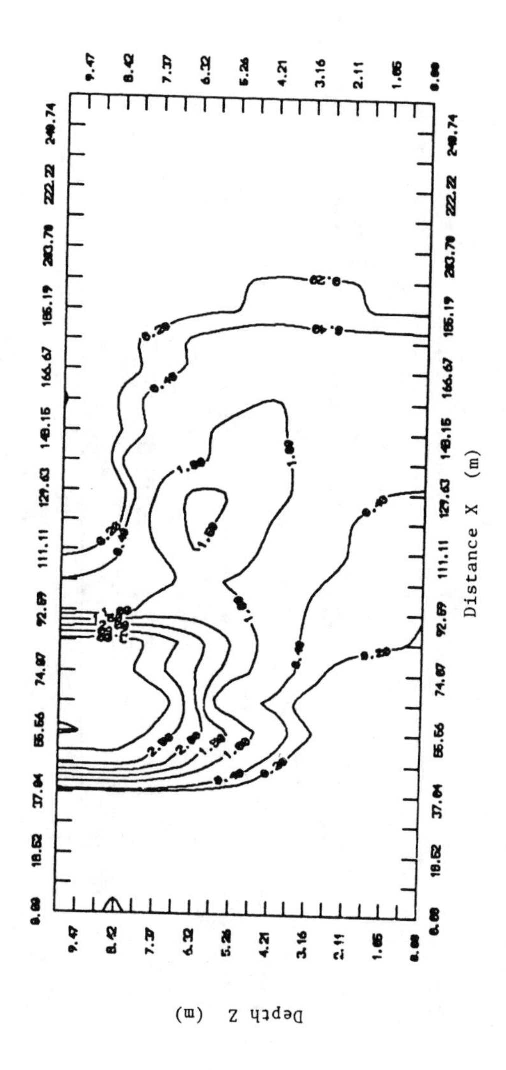


**FIG. 7, Predicted water phase concentration of Toluene after 120 days.**

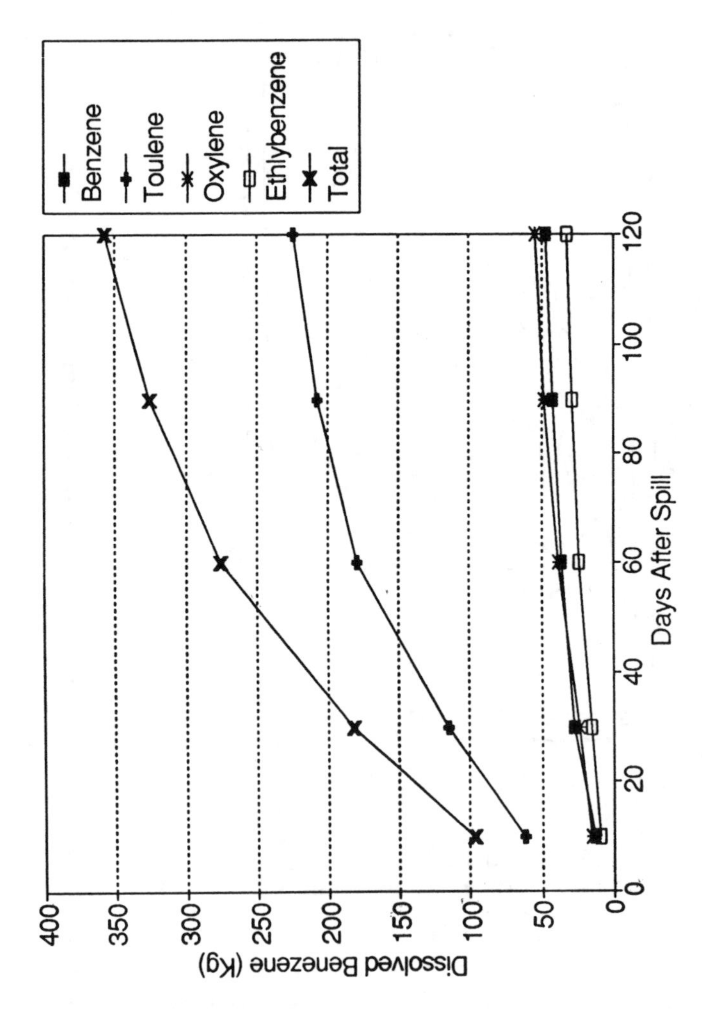


**FIG. 8, MASS of BTEX's dissolved in ground water during distribution.**

# An Assessment of Environmental Costs Associated with Crude Oil Pipeline Damage Caused by Earthquakes

Ronald T. Eguchi[1],M. ASCE
Susan D. Pelmulder[2]
Hope A. Seligson[3], A.M. ASCE

Abstract

This paper presents a methodology for assessing the risk of environmental contamination from oil pipeline leaks caused by earthquakes. Risk is measured both as volume of oil released and remediation cost. The methodology was developed for use on a regional scale and thus relies on a limited amount of input data. Monte Carlo techniques are used to simulate earthquake events, while a deterministic model is used to estimate the volume of oil released at a particular site. A library of cost models is used to estimate the contamination and resulting remediation cost based on the volume of oil released and the general site conditions. This methodology has been implemented in a computer program, OILOSS, and the results are presented as frequency of exceedance curves for volume of oil released and cost of remediation. The methodology is applied to two crude oil pipelines near the New Madrid Seismic Zone (NMSZ) and preliminary results are presented. This study is being sponsored by a grant from the National Center for Earthquake Engineering Research (NCEER).

---

[1]Vice President, EQE International, 3150 Bristol Street, Suite 350, Costa Mesa, California 92626

[2]Staff Engineer, Dames & Moore, 911 Wilshire Blvd., Suite 700, Los Angeles, California 90017

[3]Project Research Engineer, EQE International, Costa Mesa, California 92626

Introduction

The impact of lifeline failures on secondary and tertiary losses is a common concern among planners who have investigated the effects of large earthquakes on regional and national economies. Direct earthquake losses associated with the repair of damaged lifeline components may be significantly less than losses associated with an interruption of lifeline service. This type of comparison (direct vs. secondary losses) has been suggested by a recent study prepared for the Federal Emergency Management Agency (ATC-25, 1991.)

One of the most critical lifelines in the New Madrid Seismic Zone (NMSZ) is the oil transport system. Regionally, oil is carried from the south, to refineries in the midwest through a series of large underground pipeline systems. In one year, these systems deliver approximately 650 million barrels (bbls) of oil to various sites in the midwest. If this oil supply were to be disrupted or discontinued because of damage caused by a major NMSZ earthquake, it is possible that direct and secondary (i.e., business interruption) losses could exceed $0.46 billion (ATC-25, 1991.)

Many economic impacts and costs may result from earthquake-induced oil pipeline damage, including repair costs, the cost of oil spill remediation, the impact of oil supply disruption, and the impact of environmental damage. The type of loss addressed here, the remediation cost associated with the cleanup of spilled oil, was not considered in the ATC-25 study.

Historically, oil pipeline spills have occurred as a result of a number of operational causes. These causes include operator error, accidents such as digging or excavating near pipelines, and factors which may result from material defect or weakness, e.g., corrosion or weld failures. In most cases, these operational spills result in localized spillage of 1,000 bbls or less. During an earthquake, however, many pipeline locations may experience breaks or leaks resulting in spilled volumes of greater magnitude.

The purpose of this paper is to present a methodology for estimating the potential for spilled oil during an earthquake. In order to demonstrate the model developed in this study, the methodology has been applied to several oil pipelines traversing the NMSZ. The results of this study are intended to provide an additional dimension to the earthquake loss problem.

Past Earthquake Damage to Oil Pipeline Systems

Historically, oil pipeline systems have performed well in moderate to large earthquakes. Since 1960, there have been at least 35 events worldwide that have caused significant damage to some type of urban environment (Pelmulder and Eguchi, 1992). Nineteen of these recent events have been magnitude 7 or larger.

Some of the reasons why damage has been limited in oil pipeline systems include:

1. The pipeline materials that are used in oil systems tend to be of higher quality and are more ductile. These pipelines are more resistant to damage in areas of permanent distortions or displacements.

2. While underground components, including pipelines, are sensitive to permanent ground failure effects such as surface fault rupture, liquefaction and landslide, these effects usually occur in very limited areas, even during large earthquakes. Strong ground shaking effects, without ground failure, have caused little or no damage to well-designed pipelines.

3. Oil pipeline systems tend to be serial in nature. That is, these transmission systems transport oil long distances between delivery points. Unlike highly netted distribution systems, such as natural gas, oil transmission pipeline systems can be more easily located to avoid seismically hazardous areas.

There have been instances, however, when large earthquakes have caused significant damage to oil pipeline systems. In 1987, a magnitude 6.9 earthquake in Ecuador triggered massive landslides that resulted in significant damage to the Trans-Ecuadorian pipeline. The resulting loss has been termed "the largest single pipeline loss in history" (Crespo, 1987). Built in 1972, the Trans-Ecuadorian pipeline was a 26 inch X-60 grade steel pipe that carried oil 260 miles from the Ecuadorian oil fields east of the Andes to the Pacific Ocean port of Esmeraldas. 250,000 barrels per day flowed through this pipeline. Along the Coca River, 6.5 miles of the pipeline were completely destroyed. Local mud flows damaged 10 additional miles of the pipeline and severed it in at least 8 places. Five more miles were deformed with significant distortion and displacement of aboveground pipeline supports. A pipeline bridge across the Aquarico River, 30 miles west of the oil fields, was destroyed by flooding and resulted in over 2 miles of the pipeline being pulled off its supports. Lost revenue and the cost of reconstruction totaled $1 - 1.5 billion dollars (Crespo, 1987).

## Approach

The approach followed in this study was developed, in part, to be consistent with the results from other NCEER studies that are focusing on the effects of a large New Madrid earthquake. While other NCEER investigators were studying the geological or geotechnical aspects of a large New Madrid event, or the detailed performance of underground pipelines, this project team studied the socioeconomic impact of oil pipeline failures. Much of the analysis focused on the extension of previously developed models for seismic hazard assessment or pipeline vulnerability. New development in this phase of the study concentrated on remediation cost models for environmental cleanup, and the development of a model to incorporate uncertainties in soil properties along a pipeline or the extent of oil spillage in the event of an accident or leak. This methodology is currently further extended to investigate the business interruption potential of oil supply disruption.

In developing the methodology, there were three principal objectives that defined the scope of the analysis:

1. The methodology should be applicable on a broad regional level without excessive computational or input requirements,

2. The models should utilize available data, but not be dependent on site-specific information, and

3. The output of the analysis should provide a probabilistic estimate of loss level, rather than worst-case values.

These objectives, including the pragmatic approach to using regional seismic hazard data, essentially limited the choice of models used for estimating the volume of oil released in a particular event, the extent of contamination, and the associated cost of remediation to very simple models based on the generalized regional data.

The basic approach used in this study to estimate oil spill risks and remediation costs is illustrated in Figure 1. The approach uses Monte Carlo techniques to simulate earthquake events. Although the primary focus of this paper is on earthquake risks, both earthquake and normal operational risks can be assessed using this approach. In the case of seismic events, each simulation trial represents a possible outcome of the earthquake. The results of the entire simulation allow us to represent the distribution of losses that may occur from a single earthquake event. To obtain a distribution of losses caused by all earthquakes possible in the region, a series of events is simulated and the results combined through a weighting process based on the earthquake's probabilities of occurrence.

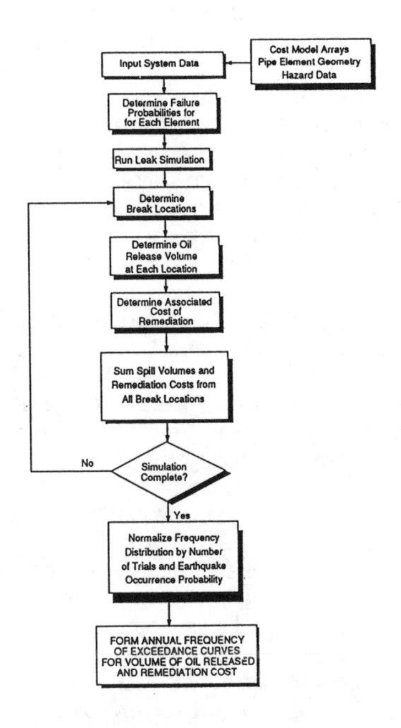


Figure 1. Procedure for Estimating Frequency of Exceedance Curves for Volume of Oil Released and Remediation Cost

A library of cost models is used to determine the remediation cost of any leak. This cost is estimated by selecting the appropriate environmental model (i.e. river crossing, flood plain, etc.) and calculating the volume of oil released. The environmental and oil release models are described in later sections of this paper.

Earthquake vulnerability models for underground pipelines developed in Eguchi (1983) and updated in Eguchi (1991) are used to determine the failure probability of each pipe segment. For each simulation trial, a random number is selected and then compared to the failure probability of a pipe element to determine if the pipe element has broken.

Once the locations of all breaks for a particular trial are known, a deterministic "drain down" model is used to estimate the volume of oil released at each location. This model assumes that all oil originally above the break drains out, with the exception of oil trapped in valleys, and oil blocked by valves and pump stations. Conservation of mass is maintained such that oil can only leak out of one location. This type of model is in some ways overly conservative because it does not consider breaks of different sizes or the time required for the pipe to drain. However, the model does provide an estimate of the volume that could leak out of the pipe if it were to break and the leak were uncontrolled. The drain down model is also easily applied in that the only data required to estimate drain down volume is the elevation at the endpoints of the pipe segments and the pipe diameter.

Once the volume released at each location is known and the appropriate environmental cost model has been specified, the remediation cost can be determined. The losses, both volume and dollar, are aggregated for all pipe failures resulting from the simulation trial to determine the total losses for the simulated event. At this point, the simulation trial is complete and a new set of break locations is generated. The process is continued until stable distributions of loss are obtained. The number of trials required is dependent on the number of pipeline elements and system failure probability. After all trials are completed, curves depicting annual frequency of exceedance are developed by normalizing the resulting frequency curves by the number of trials performed, and then unconditionalizing these curves by the annual probability of the earthquake.

## Remediation Cost Models

A library of cost models was used to determine the dollar loss associated with each simulated leak. These models estimate both the extent of environmental contamination and the cost of remediation. Five generalized models were developed to characterize the various types of contamination which may occur. The appropriate library model for each pipe segment is assigned based on the general topography of the area, depth to ground water, and proximity to surface water.

The contamination models use Darcian flow to estimate the vertical extent and rate of oil migration in soil. Using this analytical model, an estimate of the extent of ground water contamination can be made. Surface water is contaminated if the break is near a river, lake, or wetlands. Estimates of remediation costs for soil, ground water, and surface water were obtained from experts in the field. These estimates were the basis for remediation cost models, which were formulated such that the type of contamination and its associated cost could be estimated based on the library model specified and the volume of oil released.

Even with these relatively simple analytical models for contamination, there are many parameters that significantly effect the remediation cost that need to be taken into consideration, such as depth to water table, soil permeability, oil type, response time, and cost schedule. Furthermore, to be general, the models must be applicable for any season of the year and a variety of soil conditions. To obtain an estimate of the average cost for a given library model and volume released, considering these parameters, event trees were used. The parameters in the event trees were allowed to take on two or three states to approximate the range of actual values, and a probability was assigned to each state. The cost associated with the parameters in each branch of the tree was evaluated along with the likelihood of occurrence. For each model, the average or expected cost of remediation was obtained for the range of oil spill volumes specified.

While it is expected that post-earthquake response to multiple spills will have many complications, including detection difficulties, insufficient resources, and mobility and transportation problems, leading to delayed or minimized clean-up efforts and associated environmental damage, it is assumed for the purposes of this assessment that there will be no delay in implementing clean-up measures.

## Application of Methodology

This methodology was applied to two major crude oil pipelines near the NMSZ. Modified Mercalli Intensity isoseismals for a M = 8.6 earthquake within the NMSZ were developed by Algermissen and Hopper (1984). This map represents ground shaking from an earthquake occurring anywhere within the NMSZ, and is assumed to represent attenuation patterns for large earthquakes likely to occur throughout the zone. This map was digitized and used as the strong ground shaking scenario. Figure 2 shows the strong ground shaking in the study area. A single event of this magnitude ($M_S$ · 8.3) has a return period of 550 (± 125) years and 0.3 - 1.0 percent probability of occurrence by the year 2000 (Johnson and Nava, 1984). For this example, a 0.5 percent probability of occurrence by the year 2000 was assumed.

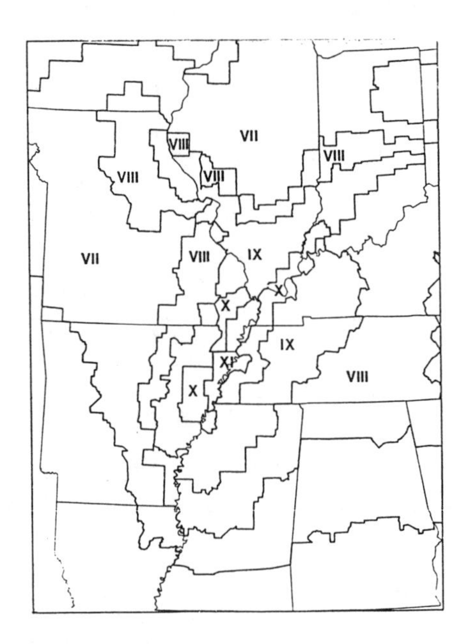


Figure 2. Modified Mercalli Intensity for a Magnitude 8.6 Event Anywhere Along the New Madrid Seismic Zone. (Digitized from Algermissen and Hopper, 1984)

Liquefaction-induced ground failure is a significant seismic hazard in the Mississippi Valley, particularly along rivers. Areas of moderate to high liquefaction potential in a large earthquake were mapped for seven states in the Central United State by Obermeier (1985). No relative distinction was made on these maps between areas of greater and lesser potential. These maps were digitized and combined into a single map. Figure 3 shows the areas of moderate to high liquefaction potential included as seismic hazard input to the simulation.

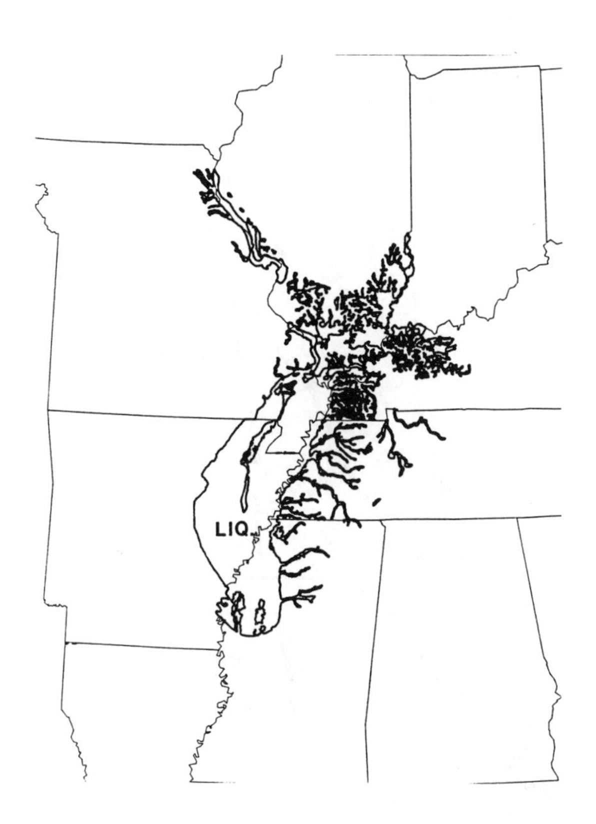


Figure 3. Areas of Moderate to High Liquefaction Potential in the New Madrid Seismic Zone During a Major Earthquake. (Digitized from Obermeier, 1985)

The Shell Capline and Mobil line number 68 were selected for use in demonstrating the methodology. These pipelines are shown in Figure 4. The Shell Capline is a 40 inch diameter pipeline built in 1968 from APR 5LX-X52 grade welded steel pipe with arc welded joints (Ariman et al., 1990). The Mobil pipeline is 20 inches in diameter in the south and 18 inches in diameter in the north. This pipeline was built around 1948 from grade B steel with welded joints.

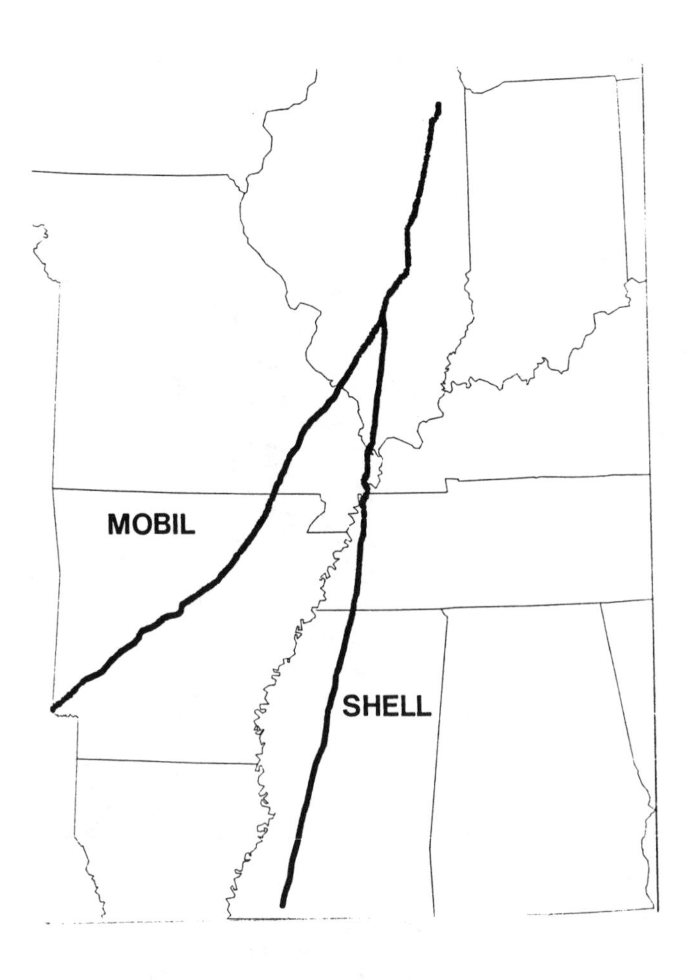


Figure 4. Pipelines Studied

The annual frequency of exceedance for released oil is shown in Figure 5. This volume is the sum of spill volumes from all leaks occurring in a single earthquake event. The range of total volumes from this scenario is 150,000 to 2,350,000 barrels, with an average spill volume of 400,000 barrels. While this would be quite a large volume for a single location, the expected number of leaks in the scenario is 87, and the expected spill volume at a single leak site is 4700 bbl. The maximum frequency on this curve, 0.005, is the frequency with which at least one spill occurs. Because this earthquake scenario represents a catastrophic event, at least one leak occurs in every simulation of the event. Therefore, the maximum frequency is also the probability of the event.

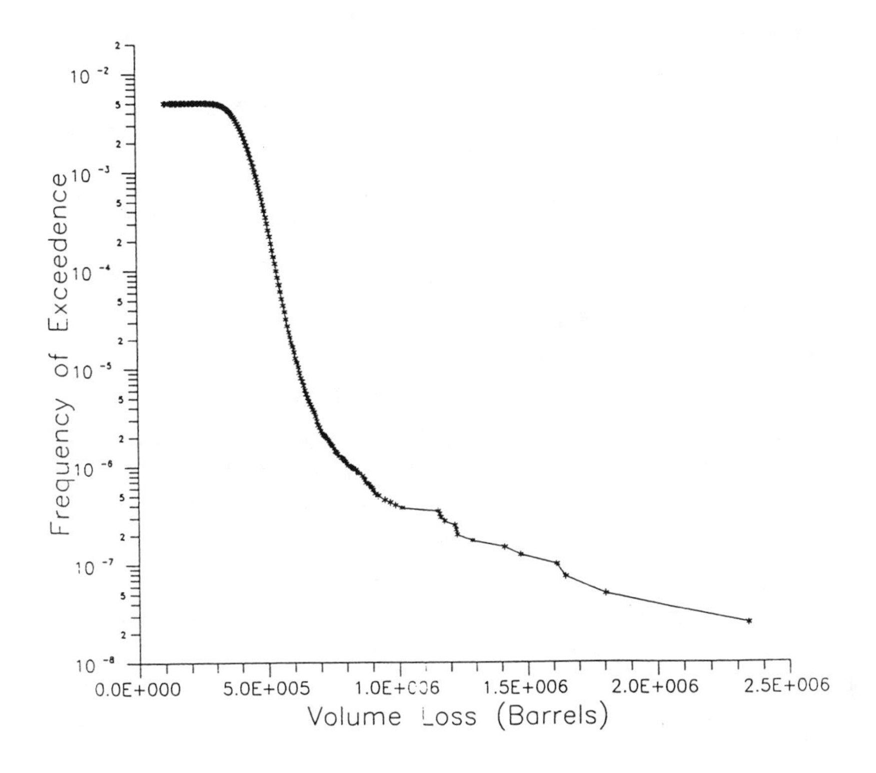


Figure 5. Frequency of Exceedance of Total Spill Volume in a Major Event in the New Madrid Seismic Zone

Dollar losses for remedial measures were also estimated for this scenario. Figure 6 presents the exceedance frequency associated with different dollar loss levels. The range of possible losses is 30 million to 2.4 billion dollars. This range is large because of the significant difference in cost models for surface water remediation and soil remediation. In a simulation where an extraordinarily large number of leaks occurs, with many of them in flood plains or rivers, the cost of clean up can be quite high. However, many more moderate cases are possible and the expected or average loss given the event is 310 million dollars. The expected loss at a single leak site is 3.6 million dollars.

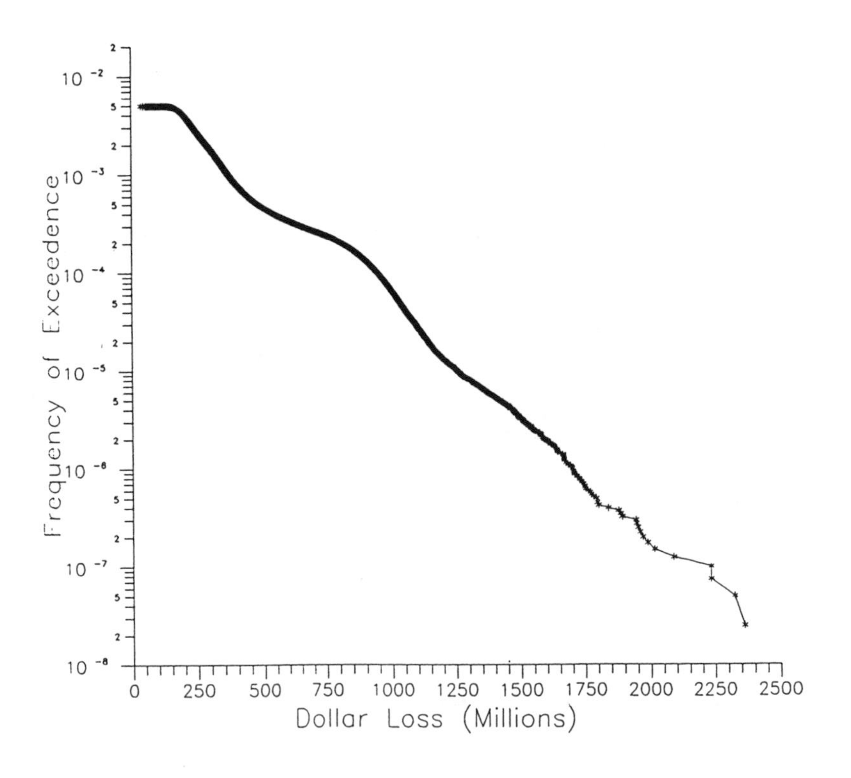


Figure 6. Frequency of Exceedance of Dollar Loss

## Conclusions

The importance of using a probability or risk-based analysis can clearly be seen from this example. If a worst case analysis had been performed, it is possible that losses in the 600,000 bbl and 2 billion dollar range would have been determined, without an estimate of the likelihood with which they would actually occur. The annual frequency of exceedance curves show that the worst case analysis may produce loss values with frequencies three to four orders of magnitude less than the frequencies of the average or expected values.

This example also demonstrates the large variation in both spill volumes and remediation costs which can occur in a given earthquake. The present methodology provides estimates of the frequency of exceedance for the entire range of volumes and costs. Using these curves, the risk associated with the operation of existing pipeline systems or the design of new systems can be quantified.

Within a broader context, the present results suggest that losses caused by environmental damage can be a significant component of the total cost of a disaster. The expected loss due to environmental cleanup of spilled oil has been estimated to be around $300 million for a large New Madrid earthquake. Repair costs and business interruption losses resulting from a disruption of oil supply have been estimated to be roughly $460 million for a similar New Madrid event (ATC-25, 1991). Therefore, an effective mitigation program designed to reduce overall losses due to oil pipeline damage must consider actions that will also minimize or isolate environmental damage.

## Acknowledgements

This research was funded by the National Center for Earthquake Engineering Research. The authors wish to thank Professor Barclay Jones of Cornell University, Professor Masanobu Shinozuka of Princeton University, and Professor Kathleen Tierney of the University of Delaware for their support in this project.

Appendix

1 bbl (barrel) = 42 U.S. gallons = 159.0 liters

1 mi = 5280 ft = 1609 m

References

Applied Technology Council (1991) Seismic Vulnerability and Impact of Disruption of Lifelines in the Conterminous United States, ATC-25.

Algermissen, S.T., and M.G. Hopper (1984) "Estimated Maximum Regional Seismic Intensities Associated with an Ensemble of Great Earthquakes that Might Occur Along the New Madrid Seismic Zone, East-Central United States," Miscellaneous Field Studies Map MF-1712, U.S. Geological Survey.

Ariman, T., et al. (1990) Pilot Study on Seismic Vulnerability of Crude Oil Transmission Systems, NCEER Technical Report NCEER-90-008.

Crespo, E., et al. (1987) "1987 Ecuador Earthquakes of March 5, 1987: EERI Special Earthquake Report," EERI Newsletter, Vol. 21, No. 7, July.

Eguchi, R.T. (1983) "Seismic Vulnerability Models for Underground Pipes," Earthquake Behavior and Safety of Oil and Gas Storage Facilities, Buried Pipelines and Equipment, PVP - Vol. 77, ASME, New York.

Eguchi, R.T. (1991) "Seismic Hazard Input for Lifeline Systems", Structural Safety, 10, Elservier Science Publishers B.V., pp. 193-198.

Helweg, O.J. (1992) "Migration of Spilled Oil from Rupture Underground Crude Oil Pipelines in the Memphis Area," Proceedings from ASCE Session on Lifeline Earthquake Engineering in the Central and Eastern United States - Energy Distribution, ASCE New York Convention.

Johnson, A. and S.J. Nava (1984) "Recurrence Rates and Probability Estimates for the New Madrid Seismic Zone," in Proceedings of the Symposium on The New Madrid Seismic Zone, P.L. Gori and W.W. Hays, eds., USGS Open File Report 84-770.

Obermeier, S.F. (1985) Plates 1-7, in Estimation of Earthquake Effects Associated with Large Earthquakes in the New Madrid Seismic Zone, M.G. Hopper, ed., U.S. Geological Survey, Open File Report 85-457.

Pelmulder, S.D. and R.T. Eguchi (1992) Seismic Risk Assessment of Crude Oil Pipelines in the Central United States, Report Prepared for National Center for Earthquake Engineering Research.

# EARTHQUAKE COUNTERMEASURES FOR LIFELINES IN THE CENTRAL AND EASTERN UNITED STATES

T.D. O'Rourke, M.ASCE[1]

Abstract

Seismic hazards affecting petroleum transmission systems in the Central U.S. and water and gas distribution systems in the Eastern U.S. are discussed. Experience during previous earthquakes is summarized for petroleum transmission pipelines, and used to delineate the main seismic and geotechnical factors which could affect the performance of petroleum transmission systems in the Midwest. Examples of earthquake countermeasures and system improvements are presented for petroleum transmission pipelines in the Central U.S., and for New York City gas and water distribution networks.

## Introduction

During an earthquake, damage to buried lifelines may be caused by traveling ground waves and permanent ground deformation. Traveling waves will cover a larger area than that in which permanent ground movements occur, and thus the opportunities for disturbing weakened or constrained portions of a piping system are correspondingly greater. Permanent ground movements, although localized, often exceed the peak ground displacements from seismic waves, and may be associated with severe and catastrophic deformation. Accordingly, they represent conditions of maximum distortion for buried pipelines and conduits, and can be taken as the most severe deformations possible during an earthquake.

Transmission pipelines are long, linear systems which usually cross many different hazard areas, each with varying degrees of predicted transient and permanent ground movement and distinct geotechnical characteristics. In the

---

[1]Professor, School of Civil and Environmental Engineering, Cornell University, Ithaca, NY 14853

Central U.S., petroleum pipelines are generally of modern construction. Their seismic vulnerability is influenced principally by the threat of large ground deformation, usually associated with soil liquefaction and landslides. Accordingly, the potential for damage to these systems depends on their proximity to a seismic source with sufficient energy to trigger liquefaction and landslides and on the location of soil deposits susceptible to such failures.

Eastern distribution systems, in contrast, involve many old and brittle components. They are vulnerable to levels of seismic intensity considerably less than those required to damage petroleum transmission lines. Even though the seismicity of many eastern cities is relatively low, the potential for damage still exists because they involve older and weaker systems. Under these conditions, traveling ground wave effects are likely to play an important role in the pattern of damage and the system-wide requirements for planning and operations.

In this paper, system improvements and earthquake countermeasures are recommended for petroleum transmission pipelines in the Central U.S. and for gas and water distribution networks in the Eastern U.S. The paper begins with a review of the performance of petroleum pipeline systems in previous earthquakes. A description of crude oil transmission pipelines in the Central U.S. is given, with discussion of the principal seismic hazards associated with the New Madrid area. Recommendations are offered for designs, construction methods, and operational policies to improve seismic performance. Finally, the gas and water distribution networks in New York City are reviewed. These systems are typical of many eastern cities. Recommendations for improving their seismic performance are provided.

## Performance During Previous Earthquakes

A review of petroleum facilities damaged during four earthquakes is presented under the following headings.

1964 Alaska Earthquake. Soil liquefaction, landslides, and tectonic subsidence affected pipeline systems during the 1964 Alaska earthquake. In Anchorage, a landslide near Ship Creek ruptured fuel tanks and flooded the area with 1,200,000 liters of diesel fuel (Eckel, 1971). A landslide near the Native Hospital destroyed a petroleum storage tank and ruptured a high pressure gas main (Hansen, 1971). There were over 200 breaks in the Anchorage gas distribution system. Liquefaction-induced flow failures at Valdez involved an estimated 75 million $m^3$ of deltaic sediments and completely destroyed the port facilities.

In Seward, substantial damage was sustained by the port faculties from waves generated by submarine landslides.

The 1964 Alaska earthquake illustrates how vulnerable coastal facilities are to ground failures and waves caused by submarine landslides. The event also shows that major pipelines tend to sustain little damage if located away from areas of large ground deformation. No damage was incurred by the 29-km crude oil line from the Kenai Peninsula to Kiikisha on the Cook Inlet, nor was there damage of the 1,100-km military multiproduct line from Haines to Fairbanks (Eckel, 1971).

1971 San Fernando Earthquake. The 1971 earthquake resulted in more than 1400 breaks in various piping systems, with nearly complete loss of water, gas, and sewage services in the City of San Fernando. The area of surface fault displacement caused by the earthquake was approximately one-half of one percent of the area affected by strong ground shaking (O'Rourke and McCaffrey, 1984). Nevertheless, approximately 25 to 50 percent of all pipeline breaks in the area of strong ground shaking occurred at or near fault crossings (O'Rourke and McCaffrey, 1984). In addition, the earthquake triggered over 1,000 landslides. Liquefaction-induced lateral spreading along the eastern and western shores of the Upper Van Norman Reservoir damaged water, gas, and petroleum transmission lines (O'Rourke and Tawfik, 1983). In general, lines of higher yield strength steels and modern welding fared well.

1976 Guatemala Earthquake. The 1976 earthquake affected petroleum facilities at the Puerto Cortes Refinery in Honduras. Damage to buildings and storage tanks was caused by seismic shaking, whereas damage to pipelines and docking facilities was caused by soil liquefaction and lateral spreading of granular fill and beach sediments (Dieckgrafe, 1976). Eight of the 12 pipelines and conduits connecting the dock and refinery were ruptured, with breaks recorded in fuel oil and gasoline pipelines. The dock shifted approximately 0.6 to 1.0 m seaward and submarine slides along an adjacent sandbar lost 1.2 m of draft for incoming ships, thus requiring dredging to reinstate the facility. A 450-mm-diameter crude oil pipeline of modern construction was deformed in tension by 0.6 m of soil movement over approximately 100 to 200 m, with no break or losses in the line.

1987 Ecuador Earthquake. Perhaps the most graphic example of pipeline damage from permanent ground movements is the destruction of the 660-mm-diameter TransEcuadorian Pipeline during the 1987 Ecuador earthquakes. Earthquake damage to approximately 40 km of the TransEcuadorian Pipeline represents the largest single pipeline loss in history. It deprived the country of 60% of its export revenues for six months, and cost roughly $850 million in lost sales and reconstruction.

The pipeline was composed of X-60 grade steel with a wall thickness of approximately 9.5 mm in the area of strong ground shaking. Most of the damaged portion of the line was supported above ground, either on "H" pile support frames or concrete pedestal foundation saddles. Seismic shaking had only a limited effect on the line pipe, whereas permanent ground deformation had a severe and extensive influence. Landslides, debris flows, and flooding caused most of the pipeline damage, and contributed to all ruptures and lost sections of line.

In summary, seismic waves develop strains in buried pipelines, but because there are little or no inertial effects from dynamic excitation, the strains tend to be low, often constituting a small fraction of the yield or rupture threshold of the pipeline material. There is only one documented case, during the 1985 Michoacan earthquake in Mexico City, for which rupturing of a modern, girth-welded steel pipeline can be shown to have been caused by seismic ground waves (O'Rourke, 1988). In contrast, there is little evidence from previous earthquakes of damage to continuous girth welded steel pipelines when construction was performed according to modern specifications and quality control procedures. There is virtually no record of damage to this type of petroleum pipeline from traveling ground wave effects.

As evinced by the case histories, the greatest earthquake damage to petroleum facilities has been caused by large ground deformations. Because the fate of most buried and large transmission pipelines is the fate of the ground, engineering measures to reduce seismic risk should focus on the hazards of ground failure.

Crude Oil Transmission Systems.

Crude oil transmission systems in the Central U.S. have been the subject of several investigations (Beavers, et al., 1986; Ariman, et al., 1990; Nyman and Hall, 1991; O'Rourke, et al., 1992). Concern has been expressed with respect to the performance of these pipelines in the event of an earthquake similar to the sequence of four main shocks, which occurred in the vicinity of New Madrid, MO in 1811-1812. Street and Nuttli (1984) estimated that three of these shocks were equivalent in surface wave magnitude, $M_S$, to a value between 8.4 and 8.7. The New Madrid earthquake sequence caused widespread liquefaction and large ground deformation in alluvial deposits of the

Mississippi River and its tributary system (Fuller, 1912; Obermeier, 1985).

Figure 1 shows the approximate locations of three major crude oil transmission pipelines, which pass close to the seismic source for earthquakes in the New Madrid area. The diameter of each pipeline is indicated in the figure. The amount of crude oil carried by these pipelines has been estimated as 1.4 million barrels per day (bbl/day), or about one-half the total amount transported to refineries in the Midwest (Beavers, et al., 1986). The most prominent of these lines, known as Capline, was the largest pipeline of its kind in the "free" world when constructed about 25 years ago. The system consists of 1000 km of line pipe and 16 pump stations. It delivers as much as 1 million bbl/day, primarily from Gulf of Mexico wells and overseas sources, to refineries in the northern Midwest.

The assumed seismic source in Figure 1 was drawn as a line connecting the epicenters of New Madrid earthquakes in 1811, 1812, 1843, and 1895. The source line was configured to be consistent with the trend of seismic activity from 1974 to 1983 reported by Johnston and Nava (1985).

Also shown in the figure are zones of liquefaction-induced ground deformation mapped by Fuller (1912) and contours of Liquefaction Severity Index (LSI). The LSI, as proposed by Youd and Perkins (1987), is defined as the maximum lateral spread displacement in inches which is likely to occur on mildly sloping, recent Holocene deposits that are susceptible to liquefaction. Attenuation relations, developed by Youd, et al. (1989) for the Eastern U.S., were used to estimate two levels of LSI, corresponding to 10 and 50. These LSIs pertain to an earthquake similar in magnitude to one of the principal 1811-1812 shocks occurring anywhere on the assumed seismic source line. Each of these LSI contours, therefore, define a zone of influence within which liquefaction-induced, lateral deformation is likely to exceed the LSI in inches for a repeat of the 1811-1812 sequence of earthquakes.

As can be seen in the figure, long sections of the petroleum pipelines fall within zones of potentially large deformation. The most likely locations of liquefaction-induced ground deformation occur at river crossings. Studies undertaken by the National Center for Earthquake Engineering Research (NCEER) have shown the Capline is the most exposed of the three pipelines to potential ground deformation (Ariman, et al., 1990).

## River Crossings

Figure 2a shows a cross-section of a typical pipeline

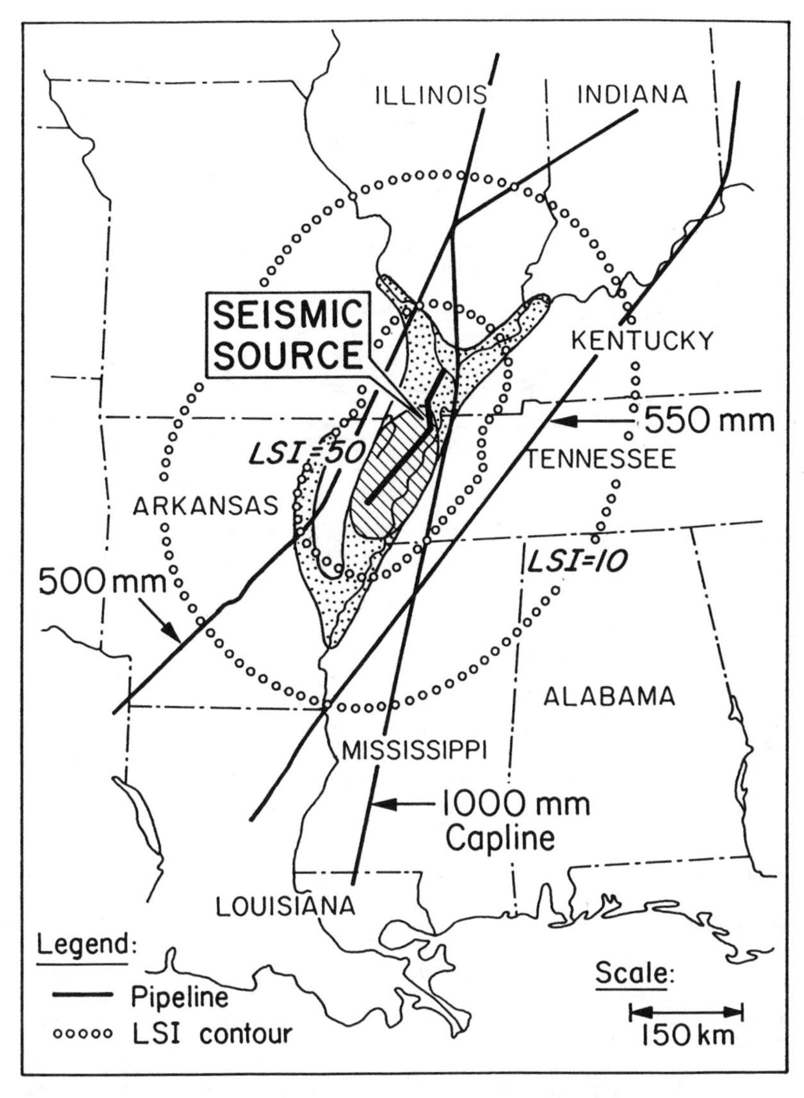


Area of principal disturbance, characterized by general warping, ejections, fissuring and severe landslides

Area of minor disturbance, characterized by local fissuring, caving of stream banks, etc.

Figure 1. Crude Oil Pipeline Systems, Zones of Ground Movement, and LSI Contours for 1811-12 Earthquake Sequence

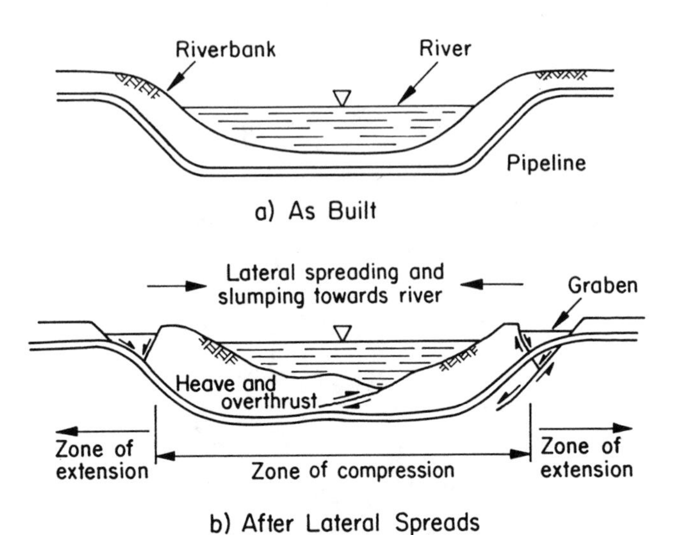


Figure 2. Profile View of Pipeline River Crossing Before and After Liquefaction-Induced Lateral Spreading

river crossing. Such crossings were generally constructed by trenching and installing the line pipe at a constant depth to a minimum set back distance (typically 5-6 m) from banks of the river. The pipeline is set at a level beneath the riverbed sufficiently deep to be protected from disturbances such as scour, future dredging, anchor dragging, etc. Bends and sloping sections are located on opposite sides of the river, where the pipeline rises in elevation to its nominal burial depth.

The river crossings are particularly vulnerable to liquefaction-induced lateral spreading. Experience in previous earthquakes has shown that such ground failure occurs as lateral displacements of one or both banks towards the river, as illustrated in Figure 2b. The pipeline is subjected to tension in the zones of lateral soil extension and to compression where it crosses beneath the river. Compression is potentially the most troublesome source of deformation because it promotes shell wrinkling of the pipe, which has been shown to be the most severe mode of distortion for girth welded steel pipe (Committee on Gas and Liquid Fuel Lifelines, 1984). The bends and sloping sections of pipe can be subject to large passive soil forces as the ground moves laterally against them. In the extreme, these sections of line may behave as anchors, thereby concentrating compressive movements within a relatively short pipe length beneath the river.

Petroleum pipeline ruptures at river crossings have obvious consequences with respect to environmental contamination.

In response to recent legislation for the protection of wetlands, much new pipeline construction in flood plains has been accomplished by means of directional drilling. Figure 3 illustrates how directional drilling is performed. Directional drilling procedures have been discussed in detail elsewhere (e.g., Hair and Shiers, 1985), and only the salient features are provided in this paper. The method involves first drilling a curved pilot hole, approximately 80 mm in diameter, along the centerline of the proposed pipeline. Bentonite drilling mud is pumped down the center of the drill rods to turn a down-hole motor just behind the drill bit. To relieve stress, a 125-mm-diameter washover pipe is drilled over the pilot string behind the drill bit. When the washover pipe exits at the reception point on the opposite side of the river, the drillstring is withdrawn and a back-reamer is attached to the washover pipe. One or two stages of back-reaming may be required, and then the actual pipeline is pulled back from the far river bank. Steering is accomplished by an instrument package (usually containing accelerometers and magnetometers) behind the down-hole motor that radios information to a computer-driven tracking station. Deviations from target coordinates are corrected by adjustments in a slightly curved section of the drill string, called a bent housing.

Directional drilling has many advantages relative to trenching and dredging with respect to seismic safety. The method is capable of installing substantial sections of pipeline, often 0.5 to 1 km long (in some cases up to 1.5 km long), and thus, the launching and reception points for the installation can be located at sufficient distances from a river to be outside the zone of liquefaction-induced lateral spreading associated with current and abandoned stream beds. Soils most favorable for directional drilling include clays, silts, and silty sands. Hence, the most advantages horizon for drilling frequently will involve soils with low or negligible liquefaction potential. Depth is usually not as constraint because there is little expense associated with drilling at deeper levels, provided that obstructions or hard rock conditions are not encountered. Accordingly, the depth of installation can be selected to circumvent liquefiable deposits.

## Recommendations for Midwestern Oil Transmission Systems

As Nyman and Hall (1991) point out, the most important issue for earthquake risk mitigation of crude oil pipelines in the Central U.S. is the adoption of a policy which deals pragmatically with the cost/benefit scenarios associated

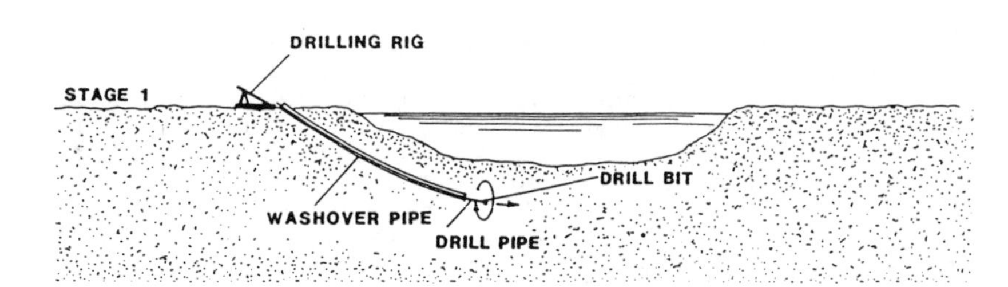


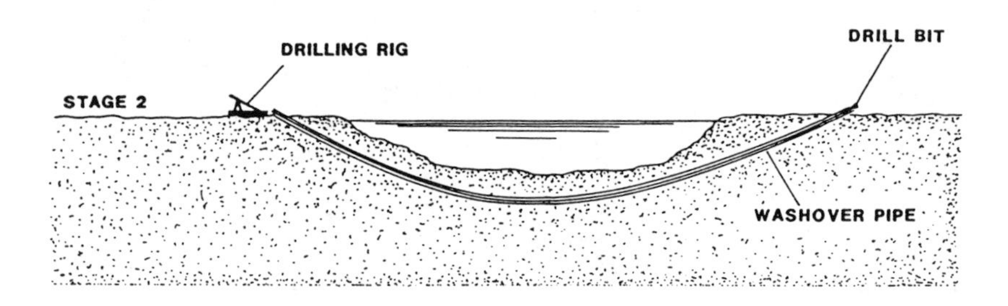


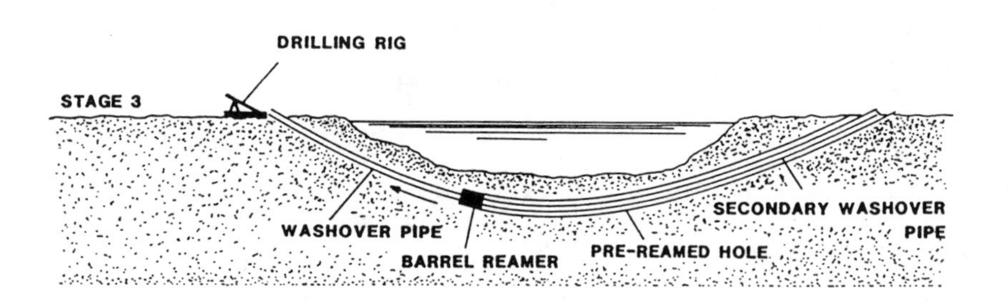


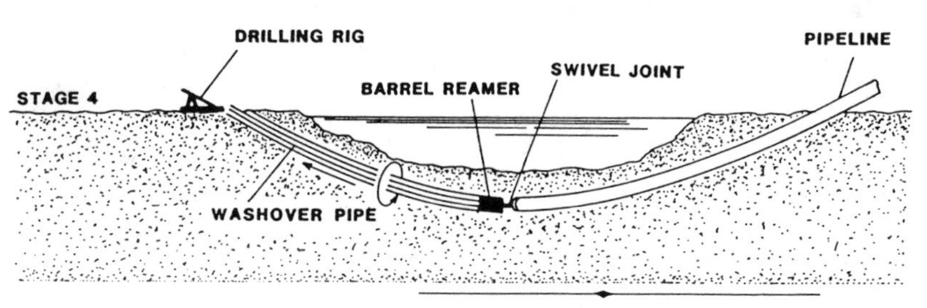


TAKEN FROM LAND & MARINE, INC. ADVERTISING BROCHURE

Figure 3. Sequence of Construction for Directional Drilling of Pipelines at River Crossings

with long and uncertain earthquake recurrence and the uncertainties involved in projecting pipeline damage. In view of the expense of improving seismic performance and the low probability of an earthquake similar to one of the main shocks of 1811-1812, it seems most appropriate to concentrate on minimizing the consequences of a severe earthquake. To this end, Nyman and Hall (1991) recommend several precautionary measures. For example, they advocate a program of line break isolation, whereby additional valves would be introduced to limit spillage at critical river crossings. Oil companies should prepare both contingency and repair plans to facilitate reinstatement and clean-up in the event of a destructive earthquake. The integrity of structures and equipment anchorages at pump and compressor stations should be checked using building code approaches, and seismic qualification should be pursued for certain critical electronic equipment.

There have been substantial advances in our understanding of liquefaction-induced ground movement and the geologic controls affecting the patterns of earthquake-induced soil movement. Accordingly, river crossings should be evaluated from a geotechnical perspective to delineate potential zones of deformation. This exercise can point out the best locations for emergency shut-off valves, and would help in developing comprehensive contingency and repair plans.

Directional drilling techniques offer advantages when installing new crossings, especially if information is available about the depth and areal disposition of liquefiable deposits. When coupled with appropriate geotechnical input, the depth of directional drilling and locations of launching and egress areas can be chosen to avoid zones of compressive ground movement, and potentially to circumvent liquefiable deposits in their entirety.

Earthquake damage is frequently concentrated at locations of restraint, such as connections with tanks, bridge abutments, and buildings, particularly when the structures are built on artificial fill or loose alluvial sediments. The risk of damage can be reduced by designing for connections that allow differential movement and permit the pipe to respond to offsets by means of slip, rotation, and ductile elongation.

## Earthquake Effects in Eastern Cities

Large ground displacements and the specific locations where such deformations are likely to occur in a relatively large magnitude event are fundamental to the process of seismic risk reduction for a transmission pipeline system in the New Madrid area. In contrast, pipeline distribution systems are more diffuse, both in the areal pattern of interconnecting lines and the variable susceptibility of these lines to earthquake damage. The old and relatively fragile pipelines in many eastern cities and the uncertainties of defining prospective zones of permanent ground deformation for regions of infrequent and relatively low magnitude earthquakes creates a situation in which risk must be assessed on a general, less site-specific basis. For many eastern cities, damage scenarios are estimated best on a global basis, using correlations between the repair rates observed in similar systems after past earthquakes and an estimate of seismic intensity, such as the Modified Mercalli Intensity (MMI). Local adjustments in the assessment of potential damage can be made in a supplementary way for areas in which ground failures are possible.

This type of process is applied in this paper for New York City (NYC). Both the NYC gas and water distribution systems are considered.

Historical accounts and instrumental readings show a relatively high frequency of small to moderate-size earthquakes in the NYC area compared with other parts of the northeastern U.S. Kafka, et al. (1985) point out that earthquake activity near NYC is broadly distributed throughout geologic features which surround a rock system of Triassic and Jurassic age, known as the Newmark Basin. Unfortunately, the geologic structures associated with most earthquakes in this region are unknown. As a result, there is considerable uncertainty about the causes, most probable locations, and potential maximum magnitudes of earthquakes which could affect NYC.

The historical record (Smith, 1966) indicates that there have been two events with maximum Modified Mercalli Intensity (MMI) of VII in the greater NYC area, in 1737 and 1884. The 1884 event was felt throughout an area of 270,000 $km^2$. Kafka, et al. (1989) suggest that the 1884 earthquake provides a minimum value for the size of earthquake which could be expected to occur near NYC, and ascribe a body wave magnitude $m_b$ = 5.0 (roughly equivalent to Richter $M_L$ 5.0) to the event. Scawthorn and Harris (1989) estimated building damage in NYC from an M 6.0 earthquake (equivalent to $M_L$ 6.0) at the epicenter of the 1884 event. In their judgment, such an earthquake represents a large, but possible, event for NYC. The earthquake intensities predicted for this scenario are shown in Figure 4 and range primarily from MM VIII near Rockaway Beach to MM VI throughout most of Brooklyn and Queens.

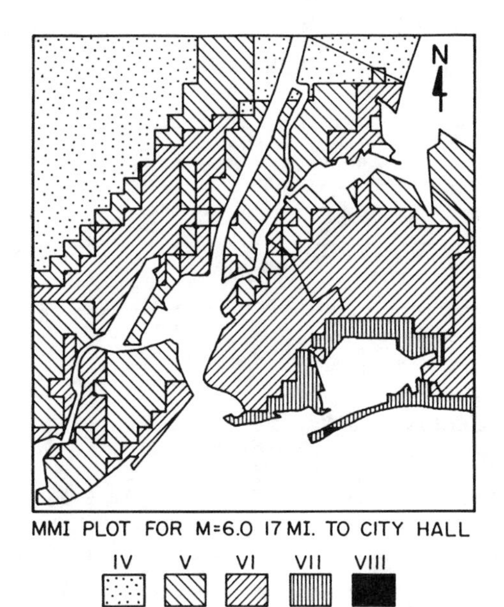


Figure 4. Map of MMI for New York City Earthquake (after Scawthorn and Harris, 1989)

These earthquake intensities would be borne by a distribution pipeline network composed predominantly of cast iron mains. Table 1 gives a breakdown of the gas distribution system in New York City (Ahmed, 1990) in 1987. About 55% of the cast iron network is composed of 150-mm-diameter pipe. Of a total of 12,885 km of mains, 6,341 km, or about 50%, are composed of cast iron. The rest are steel and plastic, comprising 46 and 4%, respectively, of the system.

Data furnished for 1985 by the New York City Bureau of Water Supply and Wastewater Collection (Ahmed, 1990) are summarized in Table 2. The data indicate that cast iron comprises 90% of the distribution network, which has a total length of 10,000 km. The two sizes which dominate the 8,422 km of cast iron pipe in the network are the 200 and 300-mm-diameter pipes. The 200 and 300-mm-diameter pipes have lengths of 4,016 and 2,848 km, respectively, and together make up 77% of the cast iron system. Apart from cast iron, the water system is composed of ductile iron, steel, galvanized wrought iron, and reinforced concrete, in percentages of 6.3, 2.7, 0.5, and 0.5%, respectively.

As mentioned previously, pipeline damage scenarios can

Table 1. Breakdown of New York City Gas Distribution System by Pipe Material (after Ahmed, 1990)

| Length by Type, km | | | |
|---|---|---|---|
| Cast Iron | Steel | Plastic | Total Length by Size |
| 6379 | 6090 | 494 | 12963 |

Table 2. Breakdown of New York City Water Distribution System by Pipe Material and Size [Source: Bureau of Water Supply and Wastewater Collection, New York City, 1985 (after Ahmed, 1990)]

| | Length of Pipe Material, km | | | | | |
|---|---|---|---|---|---|---|
| Nominal Pipe Diameter (mm) | Cast Iron | Ductile Iron[b] | Steel | Galvanized Wrought Iron | Reinforc. Concrete | Total Length by Size (km) |
| 100 | 74[a] | - | - | - | - | 74 |
| 150 | 768 | 267 | - | - | - | 1035 |
| 200 | 4016 | 222 | 2 | 2 | - | 4242 |
| 250 | 21 | - | - | 38 | - | 59 |
| 300 | 2848 | 74 | 14 | 3 | - | 2939 |
| 350 | 8 | - | - | 6 | - | 14 |
| 400 | 235 | - | 2 | - | - | 237 |
| 500 | 718 | - | 8 | - | - | 726 |
| 600 | 69 | - | 2 | - | - | 71 |
| 750 | 66 | - | 8 | - | - | 74 |
| 900 | 93 | - | 26 | - | 10 | 129 |
| 1050 | 3 | - | - | - | - | 3 |
| 1200 | 154 | - | 98 | - | 37 | 289 |
| 1370 | - | - | - | - | - | - |
| 1520 | - | - | 45 | - | 2 | 47 |
| 1670 | - | - | 18 | - | - | 18 |
| 1830 | - | - | 51 | - | 3 | 54 |
| 2130 | - | - | - | - | 2 | 2 |
| Total Length by Matl. | 9073 | 563 | 274 | 49 | 54 | 10013 |

[a] - combined cast and ductile iron
[b] - estimated on basis of replacement trends (Betz, Converse, Murdoch; 1980)

be estimated on the basis of repair statistics from previous earthquakes which have been correlated with MMI. Table 3 summarizes repair statistics for cast iron water pipeline systems damaged by various U.S. earthquakes. Table 4 summarizes similar statistics compiled by O'Rourke, et al. (1991) for the Municipal Water Supply System (MWSS) in San Francisco as a result of the 1989 Loma Prieta earthquake. The preponderance of MWSS pipelines are composed of cast iron. Table 3 reflects data acquired for portions of cast iron pipeline systems influenced predominantly by ground shaking. For example, the 1906 San Francisco statistics have been screened of pipeline breaks in areas of liquefaction and lateral spreads as mapped by O'Rourke, et al. (1992). A similar screening process was performed by Eguchi (1982) for the 1971 San Fernando earthquake. In contrast, Table 4 shows repair statistics for areas of MMIX to MMVII in which there was clear evidence of liquefaction. A repair indicates a location where excavation and corrective action were recorded by utility crews. Because of limitations in the utility records, however, it is not possible to identify the specific cause of damage for many of the recorded repairs.

Repair statistics from Tables 3 and 4 are plotted in Figure 5. Because of the difficulties in discriminating pipeline damage caused by permanent relative to transient ground deformation, the data from the 1906 San Francisco earthquake and from MMIX and MMX areas of the 1971 San Fernando earthquake are not included in the plot. Coefficients of determination, $r^2$, associated with linear regressions of the 1989 Loma Prieta data and all combined data are shown. The slopes of both regressions are virtually identical. These slopes indicate that there is nearly an order of magnitude increase in repair rate for each increment of MMI.

The figure shows considerable variability in the data. This variability may be associated with different states of repair. For example, damage statistics reported for the 1983 Coalinga earthquake (Isenberg and Taylor, 1984) suggest that there was a relatively high rate of corrosion in the system. Damage in San Francisco may be related in various areas to residual pipeline stresses caused by consolidation and settlement of underlying Holocene bay mud. Perhaps the greatest source of variability is associated with uncertainties in the assignment of MMI. The MMI designation is judgmental, thereby contributing to variability in the actual number assigned. Moreover, the spatial extent of a given MMI is inherently uncertain and directly affects the estimation of pipeline km associated with a given number of repairs. The choice of affected

Table 3. Summary of Cast Iron Pipeline Damage Statistics Related to Transient Ground Motion During U.S. Earthquakes

| Earthquake & References | Damage Observ. | Nominal Pipe Diameter, mm | | | | | Totals | Intensity[a] |
|---|---|---|---|---|---|---|---|---|
| | | 75-150 | 200-250 | 300-400 | 450-1050 | > 1050 | | |
| 1906 San Francisco | Repairs | - | - | - | - | - | 135 | VIII |
| | Length (km) | - | - | - | - | - | 644 | |
| | Ratio | - | - | - | - | - | 0.210 | |
| 1949 Seattle (Kennedy/ Jenks/ Chilton, 1990) | Repairs | - | 16[b] | - | 1[c] | - | 17 | VII |
| | Length (km) | - | 1176 | - | 143 | - | 1319 | |
| | Ratio | - | 0.013 | - | 0.007 | - | 0.013 | |
| | Repairs | - | 18[b] | - | 6[c] | - | 24 | VIII |
| | Length (km) | - | 53 | - | 32 | - | 85 | |
| | Ratio | - | 0.342 | - | 0.190 | - | 0.283 | |
| 1965 Puget Sound (Kennedy/ Jenks/ Chilton, 1990) | Repairs | - | 11[b] | - | 3[c] | - | 14 | VII |
| | Length (km) | - | 1718 | - | 189 | - | 1907 | |
| | Ratio | - | 0.006 | - | 0.016 | - | 0.007 | |
| | Repairs | - | 10[b] | - | 3[c] | - | 13 | VIII |
| | Length (km) | - | 94 | - | 18 | - | 112 | |
| | Ratio | - | 0.106 | - | 0.166 | - | 0.116 | |
| 1969 Santa Rosa (Isenberg, 1978a & 1978b; Bigglestone, 1970) | Repairs | 1 | 1 | 5 | - | - | 7 | VII |
| | Length (km) | 145 | 43 | 31 | - | - | 219 | |
| | Ratio | 0.007 | 0.023 | 0.161 | - | - | 0.032 | |
| 1971 San Fernando (Isenberg, 1978a, 1978b & 1979; Eguchi, 1982) | Repairs | 15 | 0 | 1 | - | - | 16 | VII |
| | Length (km) | 450 | 173 | 43 | - | - | 666 | |
| | Ratio | 0.033 | 0 | 0.023 | - | - | 0.024 | |
| | Repairs | 15 | 19 | 2 | 3 | - | 39 | VIII |
| | Length (km) | 150 | 70 | 30 | 7 | - | 257 | |
| | Ratio | 0.100 | 0.271 | 0.067 | 0.429 | - | 0.152 | |
| | Repairs | 43 | 23 | 15 | 3 | - | 84 | IX |
| | Length (km) | 340 | 141 | 50 | 5 | - | 536 | |
| | Ratio | 0.126 | 0.163 | 0.300 | 0.600 | - | 0.157 | |
| | Repairs | 5 | 3 | 1 | 0 | - | 9 | X |
| | Length (km) | 10 | 7 | 3 | LT | - | 20 | |
| | Ratio | 0.500 | 0.429 | 0.333 | 0 | - | 0.450 | |
| 1983 Coalinga (Isenberg & Taylor, 1984) | Repairs | 10 | 1 | - | - | - | 11 | VIII |
| | Length (km) | 12 | 2 | - | - | - | 14 | |
| | Ratio | 0.833 | 0.500 | - | - | - | 0.786 | |
| 1987 Whittier (Wang, 1990) | Repairs | 8 | 3 | - | - | - | 11 | |
| | Length (km) | 28.4 | 7.8 | - | - | - | 36.2 | |
| | Ratio | 0.28 | 0.38 | - | - | - | 0.30 | |

Notes: a - refers to Modified Mercalli Intensity; LT - less than 0.5 km
b - refers to diameter $\leq$ 300 mm; c - refers to diameter > 300 mm

Table 4. Summary of MWSS Pipeline Repairs for Various Sites in San Francisco after the 1989 Loma Prieta Earthquake (O'Rourke, et al., 1991)

| Site | Repairs | Length, km | Repairs per km | Modified Mercalli Intensity, MMI |
|---|---|---|---|---|
| Marina | 69 | 11.3 | 6.11 | IX |
| South of Market | 13 | 17.1 | 0.76 | VIII |
| Foot of Market | 6 | 13.8 | 0.43 | VII |
| Mission Creek | 2 | 15.1 | 0.13 | VII |
| Remainder of System | 15 | 1740 | 0.01 | VI |

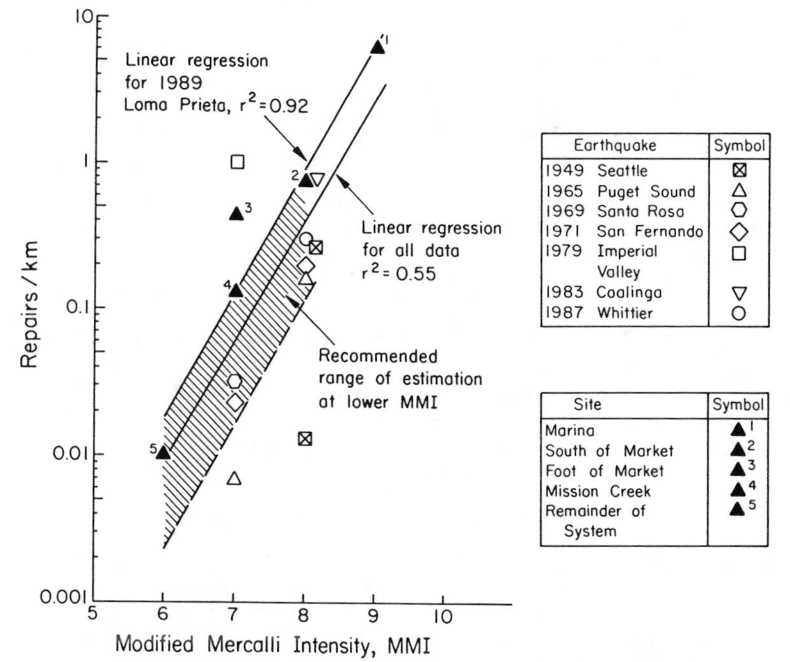


Figure 5. Repair Rate of Cast Iron Pipeline Systems Versus Modified Mercalli Intensity (adapted from O'Rourke, et al., 1991)

area and corresponding pipeline km can have a strong influence on repair rates. For example, Wang (1990) analyzed cast iron pipeline repair data for the entire City of Whittier and for only a central zone in Whittier. Although both areas were characterized by MMVIII, the central zone showed 0.30 repairs/km, nearly four times larger than 0.08 repairs/km for the entire city.

The trends shown in Figure 5 can be used to estimate damage in the cast iron pipeline networks of NYC. A recommended range of estimation is shown for lower MMI, which indicates that empirical correlations will yield repair rates that vary by nearly an order of magnitude. It should be recognized that the upper bound regression, based on Loma Prieta earthquake data, is representative of soft soil sites and areas prone to liquefaction. The lower bound trend may be applicable for areas where site amplification and permanent ground movements are unlikely. Using Figures 4 and 5, it can be seen that distribution damage in NYC would vary significantly, depending on local MMI. For the upper bound regression in Figure 5 associated with MM VII to VIII, earthquake damage to the pipeline network would overwhelm local and district crews. Given the earthquake scenario of Scawthorn and Harris (1989), there could be scores of repairs required in the water distribution system throughout the southern portions of Brooklyn and Queens. For the gas distribution system in these areas, there could be dozens of repairs needed.

The MMI projected by Scawthorn and Harris (1989) for most of NYC is less than or equal to VI. Average distribution system damage for these projections is low. Accordingly, there would be the opportunity to transfer repair crews and emergency personnel from relatively undamaged sections to those with difficulties, provided that an emergency plan has been established.

It should be recognized that pipeline system damage, estimated from average repair statistics and approximate regional intensities, may not account adequately for locally severe effects associated with permanent ground deformation, or the failure of large diameter conduits which are critical for system supply. It is well known that consolidation of loose fills and the liquefaction of granular waterfront fills and loose natural soils can result in significant vertical and lateral displacement, thereby causing concentrated pipeline damage. Moreover, the pipeline system in NYC contains many old and relatively fragile components. A rupture of a 600 to 900-mm-diameter, high pressure water main can cause severe flooding and undermining of roads and adjacent utilities. There have been several water main failures of this type in NYC in recent years, where flooding, erosion, and associated disruptions have resulted in multi-million dollar losses.

There is a close relationship between damage to the water distribution system and the control of fires following an earthquake. Scawthorn and Harris (1989) estimate that as many as 130 ignitions could be caused by a M. 6.0 earthquake near NYC. They also note that the NYC Fire Department has over 200 engines, so there is substantial capacity in the city to respond to multiple fires. The availability of water also must be considered, and this depends on the damage state of the water supply system. The failure of large diameter mains can have a serious impact on pressure and flow in adjacent parts of the system.

Recommendations for NYC Distribution Networks

Given the uncertainties associated with the occurrence of earthquakes in and near NYC and the complexity of its utility systems, it is necessary to approach seismic response in a way that not only is prudent, but is sensible and pragmatic for the institutions which would implement procedures to reduce seismic risk. The most advantageous approach for utility systems is to use an established framework of operation in conjunction with the overall goals of rehabilitating the infrastructure for reliable and cost-effective service under normal conditions. In this paper, a three-fold approach is proposed involving: 1) an emergency plan, 2) replacement strategy, and 3) selective deployment of special components.

Emergency Plan. The rapid and systematic deployment of repair crews in sufficient numbers to isolate and reinstate many different locations of damage is of critical importance in an emergency such as an earthquake. As discussed in this paper, seismic damage to water supply and gas distribution systems is most likely to be concentrated in specific sections of the systems, thereby providing an opportunity for repair and emergency personnel to be sent from relatively undamaged sectors. Coordination of such a large working force requires advanced planning. Mutual assistance agreements can be drawn up among neighboring utilities, so that additional personnel and access to a greater number of emergency stockpiles would be expedited.

New York State law requires that all public drinking water supplies have emergency plans (New York State, 1990). This statutory requirement was implemented recently, primarily as a safeguard against drought and reservoir shortages. Nevertheless, it provides a vehicle for response under earthquake conditions. Water authorities and commissions in New York State must file written emergency plans to comply with recent law. Comparatively minor adjustments and additions could be undertaken to rationalize seismic response and establish automatic procedures for action in the event of an earthquake.

Replacement Strategy. Seismic risk for eastern utilities should be viewed in the broader context of infrastructure rehabilitation. It has been well established that maintenance and replacement procedures, which reduce poor performance from corrosion and mechanically-induced stresses, also will improve seismic performance. Hence, it makes sense to treat improved seismic response as a subset of maintenance and renovation. Such a strategy may provide opportunities for grants in aid under special circumstances, and will help target potential bad actors and difficult portions of the system for early replacement.

Candidates for early replacement include welded steel pipelines that were constructed before the early 1930s. Such lines generally did not benefit from the quality control of welding exercised in more recent times. They have been shown to be extremely vulnerable to seismic effects (O'Rourke and McCaffrey, 1984), and also are prone to damage at welds during severe winters. Old water mains and pipeline segments with prior histories of recurrent repair also are susceptible to continuing difficulties (Betz, Converse, and Murdoch, 1980). A rational and systematic replacement process should be adopted which includes consideration of seismic effects. The most critical lines for maintaining pressure and flow should be identified, in conjunction with an assessment of their state of repair, local soil conditions, and potential for transient and permanent ground deformation.

Selective Deployment. Cast iron mains often are replaced with ductile iron pipes utilizing flexible gasket joints. The ductile and flexible nature of these pipes enhances performance and virtually eliminates brittle fracture. In some cases, additional features should be considered. For example, special restrained joints are being used by the San Francisco Water Distribution Department in their ductile iron pipe. These commercially available joints are equipped with stainless steel inserts which engage the barrel of the spigot and bear against the bell of the adjacent pipe to form a joint which is flexible in rotation but restrained from longitudinal pullout. The restraining inserts add only a very small percentage to the cost of the pipe. Such longitudinally restrained pipelines could be deployed selectively in areas which may be influenced by permanent lateral and vertical ground movements, as well as locally strong shaking. Sites which would benefit from the use of such pipe include areas of loose waterfront fills and areas underlain by relatively deep deposits of soft to medium clay, representative of $S_4$ type soils in currently proposed building codes (e.g., Building Seismic Safety Council, 1988).

## Conclusions

For petroleum transmission systems in the Midwest, it is most appropriate to concentrate on minimizing the consequences of a severe earthquake. Nyman and Hall (1991) have recommended an overall program involving line break isolation, contingency plans, repair plans, building code checks on pump and compressor stations, and seismic qualification of critical electronic equipment.

Directional drilling techniques offer advantages when installing new crossings, especially if information is available about the depth and areal disposition of liquefiable deposits. When coupled with appropriate geotechnical input, the depth of directional drilling and locations of launching and egress areas can be chosen to avoid zones of compressive ground movement, and potentially to circumvent liquefiable deposits in their entirety.

For an earthquake of M 6.0 with an epicenter at the location of the 1884 earthquake near Rockaway Beach, substantial damage is likely in the water and gas distribution systems of NYC. Scores of repairs in the water supply would be needed in the southern portions of Brooklyn and Queens, overwhelming local and district crews. Damage in other parts of the city would be light, thereby providing an opportunity to send repair and emergency personnel from other sectors of the system.

A three-fold approach for NYC distribution systems is proposed which involves: 1) developing an emergency plan which covers seismic effects, 2) defining a strategy for annual pipeline replacements based on the relative vulnerability of system components, and 3) selective deployment of pipelines with specially restrained joints in areas with the potential for large transient and permanent ground deformation. Seismic risk for eastern utilities should be viewed within the broader context of infrastructure rehabilitation, with primary emphasis given to those measures which improve normal operation. Maintenance practices which cut down on corrosion and remove pipelines with recurrent problems will reduce seismic risk, and are an effective means of enhancing system reliability under all types of emergency operation.

Acknowledgments

This paper was prepared as part of research supported by the National Center for Earthquake Engineering Research at Buffalo, NY under Grant No. 913541A. Portions of this paper on NYC are available in the lecture notes on Geotechnical Aspects of Seismic Design in the New York Metropolitan Area, which were sponsored and distributed through the ASCE NYC Metropolitan Section. The excellent typing and drafting skills of L. Mayes and A. Avcisoy, respectively, are deeply appreciated.

References

Ahmed, I., "Pipeline Response to Excavation-Induced Ground Movements", Ph.D. Thesis, Cornell University, Ithaca, NY Aug. 1990.

Ariman, T., R. Dobry, M. Grigoriu, F. Kozen, M. O'Rourke, T. O'Rourke, and M. Shinozuka, 1990, "Pilot Study on Seismic Vulnerability of Crude Oil Transmission Systems", Technical Report NCEER-90-0008, National Center for Earthquake Engineering Research, Buffalo, NY, 1990.

Beavers, J.E., R.G. Domer, R.J. Hunt, and R.M. Torry, "Vulnerability of Energy Distribution Systems to an Earthquake in the Eastern United States - An Overview", American Association of Engineering Societies, Dec. 1986.

Betz, Converse, Murdoch, Inc., "New York District Corps of Engineers New York City Water Supply Infrastructure Study", Betz, Converse, Murdoch, Inc., Plymouth Meeting, PA, May 1980.

Bigglestone, H., "Public Services", Part III of The Santa Rosa California Earthquake of Oct. 1, 1969, U.S. Dept. of Commerce, Rockville, MD, 1970, pp. 37-44.

Building Seismic Safety Council, "NEHRP Recommended Provisions for the Development of Seismic Regulations for New Buildings", Parts 1 and 2, Building Seismic Safety Council, Washington, DC, 1992.

Committee on Gas and Liquid Fuel Lifelines, "Guidelines for the Seismic Design of Oil and Gas Pipeline Systems", ASCE, New York, NY, 1984.

Dieckgrafe, R.E., Personal Communication, transmitted from the Manager, Texaco Puerto Cortes Refinery, Honduras, Nov. 1976.

Eckel, E.B., "Effects on Air and Water Transport, Communications, and Utility Systems", The Great Alaska Earthquake of 1964, National Academy of Sciences, Washington, DC, 1971, pp. 705-731.

Eguchi, R.T., "Earthquake Performance of Water Supply Components during the 1971 San Fernando Earthquake", Technical Report No. 82-1396-2a, J.H. Wiggins Co., Rendondo Beach, CA, Mar. 1982.

Fuller, M.L., "The New Madrid Earthquakes", Bulletin 494, U.S. Geological Survey, Washington, DC, 1912.

Hair, J.D. and G.E. Shiers, "Directionally-Controlled Drilling for Pipelines", Proceedings, No-Dig 85 Trenchless Construction for Utilities, Institution of Public Health Engineers, London, U.K., Apr. 1985, pp. 160-169.

Hansen, W.R., "Effects at Anchorage", The Great Alaska Earthquake of 1964, National Academy of Sciences, Washington, DC, 1971, pp. 289-358.

Isenberg, J., "The Role of Corrosion in the Seismic Performance of Buried Steel Pipelines in Three United States Earthquakes", Grant Report No. 6, Weidlinger Associates, New York, NY, June 1978a.

Isenberg, J., "Seismic Performance of Underground Water Pipelines in the Southeast San Fernando Valley in the 1971 San Fernando Earthquake", Grant Report No. 8, Weidlinger Associates, New York, NY, Sept. 1978b.

Isenberg, J., "Role of Corrosion in Water Pipeline Performance in Three U.S. Earthquakes", Proceedings, 2nd U.S. National Conference on Earthquake Engineering, Stanford, CA, Aug. 1979, pp. 683-692.

Isenberg, J. and Taylor, C.E., "Performance of Water and Sewer Lifelines in the May 2, 1983 Coalinga California Earthquake", Lifeline Earthquake Engineering: Performance, Design, and Construction, ASCE, New York, NY, Oct. 1984, pp. 176-189.

Johnston, A.C., and S.J. Nava, "Recurrence Rates and Probability Estimates for the New Madrid Seismic Zone", Journal of Geophysical and Research, Vol. 90, 1985, pp. 6737-6753.

Kafka, A.L., E.A. Schlesinger-Miller, and N.L. Barstow, "The Earthquake Activity in the Greater New York City Area: Magnitudes, Seismicity, and Geologic Structures", Bulletin of the Seismological Society of America, Vol. 75, No. 5,

Oct. 1985, pp. 1285-1300.

Kafka, A.L., M.A. Winslow, and N.L. Barstow, "Earthquake Activity in the Greater New York City Area: A Fault-finder's Guide", Field Trip Guidebook, New York State Geological Association, 61st Annual Meeting, Oct. 1989, pp. 177-197.

Kennedy/Jenks/Chilton, "Earthquake Loss Estimation Modeling of the Seattle Water System", Report No. K/J/C 886005.00, Kennedy/Jenks/Chilton, Federal Way, WA, Oct. 1990.

Nyman, D.J. and W.J. Hall, "Scenario for Improving the Seismic Response of Pipelines in the Central United States", Proceedings, Third U.S. Conference on Lifeline Earthquake Engineering, M.A. Cassaro, Ed., ASCE, New York, NY, Aug. 1991, pp. 196-205.

Obermeier, S.F., "Nature of Liquefaction and Landslides in the New Madrid Earthquake Region", in Hopper, M.G., Ed., Estimation of Earthquake Effects Associated with large Earthquakes in the New Madrid Seismic Zone, U.S. Geological Survey, Open-File Report 85-357, 1985, pp. 34-55.

O'Rourke. M.J., M. Shinozuka, T. Ariman, R. Dobry, M. Grigoriu, F. Kozen, and T.D. O'Rourke, "Study of Crude Oil Transmission System Seismic Vulnerability", Earthquake Spectra, Earthquake Engineering Research Institute (in press).

O'Rourke, M.J., "Seismic Analysis Procedures for Welded Steel Pipeline", Proceedings, First Japan-U.S. Workshop on Liquefaction, Large Ground Deformation, and Their Effects on Lifeline Facilities, Tokyo, Japan, Nov. 1988, pp. 133-142.

O'Rourke, T.D. and M.A. McCaffrey, "Buried Pipeline Response to Permanent Earthquake Ground Movements", Proceedings, Eighth World Conference on Earthquake Engineering, San Francisco, CA, Vol. VII, July 1984, pp. 215-222.

O'Rourke, T.D. and M.S. Tawfik, "Effects of Lateral Spreading on Buried Pipelines during the 1971 San Fernando Earthquake", Earthquake Behavior and Safety of Oil and Gas Storage Facilities, Buried Pipelines, and Equipment, ASME, PVP-Vol. 77, 1983, pp. 124-132.

O'Rourke, T.D., T.E. Gowdy, H.E. Stewart, and J.W. Pease, "Lifeline and Geotechnical Aspects of the 1989 Loma Prieta Earthquake", Proceedings, 2nd International Conference on Recent Advances in Geotechnical Earthquake Engineering and

Soil Dynamics, St. Louis, MO, Vol. II, 1991.

O'Rourke, T.D., P.A. Beaujon, and C.R. Scawthorn, "Large Ground Deformations and Their Effects on Lifeline Facilities: 1906 San Francisco Earthquake", Chapter in Case Studies of Liquefaction and Lifeline Performance During Past Earthquakes, Vol. 2, Technical Report NCEER-92-0002, National Center for Earthquake Engineering Research, Buffalo, NY, Feb. 1992.

Scawthorn, C. and S.K. Harris, "Estimation of Earthquake Losses for a Large Eastern Urban Center: Scenario Events for New York City", Earthquake Hazards and the Design of Constructed Facilities in the Eastern United States, K.H. Jacob and C.J. Turkstra, Eds., Annals of the New York Academy of Sciences, Vol. 58, 1989, pp. 435-451.

Smith, W.E.T., "Earthquakes of Eastern Canada and Adjacent Areas, 1928-1959", Publication of the Dominion Observatory, Ottawa, Canada, Vol. 32, 1966, pp. 87-121.

Street, R. and O.W. Nuttli, "The Central Mississippi River Valley Earthquake of 1811-1812", Missouri Academy of Sciences Meeting, Cape Girardeau, Missouri, April 1984.

Wang, L. R-L., "A New Look Into the Performance of Water Pipeline Systems form 1987 Whittier Narrows, California Earthquake", Technical Report No. ODU LEEE-05, Old Dominion University, Norfolk, VA, Jan. 1990.

Youd, T.L. and D.M. Perkins, "Mapping of Liquefaction Severity Index", Journal of the Geotechnical Division, ASCE, New York, NY, Vol. 113, No. 11, 1987, pp. 1374-1392.

Youd, T.L., D.M. Perkins, and W.G. Turner, "Liquefaction Severity Index Attenuation for the Eastern United States", Proceedings, Second U.S.-Japan Workshop on Liquefaction, Large Ground Deformation, and Their Effects on Lifeline Facilities, National Center for Earthquake Engineering Research, Buffalo, NY, 1989, pp. 438-452.

## SUBJECT INDEX

Page number refers to first page of paper.

## AUTHOR INDEX

Page number refers to first page of paper.